INTEGRITY TESTING FOR LOW-PRESSURE MEMBRANES

Charles Liu

American Water Works Association

AWWA Publications Manager: Gay Porter De Nileon
AWWA Technical Editor/Project Manager: Melissa Valentine
Cover Art: Cheryl Armstrong
Production Editor: Sheryl Tongue

Library of Congress Cataloging-in-Publication Data

Liu, Charles, 1957-
Integrity testing for low-pressure membranes / by Charles Liu.
p. cm.
Includes bibliographical references.
ISBN 978-1-58321-793-1
1. Water--Purification--Membrane filtration. 2. Drinking water--Purification. I. Title.
TD442.5.L58 2011
628.1'64--dc23

2011034782

ISBN 1-58321-793-2
978-1-58321-793-1

Printed in the United States of America

American Water Works Association

6666 West Quincy Avenue
Denver, CO 80235-3098
303.794.7711
www.awwa.org

Dedication

To my wife, Jianying, for her constant support and encouragement through my career; and to my children, Hervey and Helen, my source of joy (and occasionally anxiety).

Contents

Preface

In the fall of 1997, shortly after joining the Pall Corporation, the author was summoned to the office of Paul Johnson. Paul asked if the author knew anything about membrane integrity. The author replied that he knew nothing but the basic concept. Paul then asked the author to lead in the effort to develop integrity testing procedures for membrane systems in the municipal water market. Paul said that he believed membrane integrity would be a very important issue in drinking water and to develop the process would be the best way to learn membrane integrity.

Paul Johnson, a former US Navy officer and a vice president of Scientific and Laboratory Services of Pall Corporation, was the type of the leader who commanded the respect and trust of his troops, gave them necessary resources, and rallied them to accomplish the mission. The meeting in Paul's office that day marked the beginning of a decade-long journey of learning about membrane integrity.

Membrane integrity is in an interesting combination of science, engineering, and regulations. It is a classic example of science and regulations interacting and coevolving in an intriguing way. In the last decade, the author was fortunate to have the opportunity to participate in three significant developments in membrane integrity testing: (1) the development of integrity testing practices for membrane systems in drinking water by the American Society of Testing and Materials (ASTM D-6908-03) in 2003; (2) the project titled *Assessment and Development of Low-Pressure Integrity Monitoring Tools* by the Awwa Research Foundation (now known as the Water Research Foundation) in 2004; and (3) and the publication of the *Membrane Filtration Guidance Manual* by USEPA in 2005. The author has gradually gained the understanding of issues and sciences on this subject by participating in these activities.

This book is an attempt to introduce the reader to the subject of integrity testing of low-pressure membranes in the context of drinking water filtration. Both theoretical and practical aspects of membrane integrity testing are discussed. In general, the discussion for each integrity-testing method includes (1) what it is, (2) how it works, and (3) how it is applied. The hope is that by including underlying principles of the methods, the reader will have a better understanding of the pros and cons of each method. Certain portions of the book involve mathematical derivation of equations (e.g., chapter 5). However, it is the author's intention to use the equations to describe the physical concepts not the mathematics per se.

In spite of the long history of membrane integrity testing, the field is still dynamic and evolving. Therefore, parts of the book will be "history" already, even with the best efforts of the author to keep the information up to date. For that reason, the author adapted a more open structure by including a chapter to describe two future technologies for membrane integrity testing.

Acknowledgments

The author is indebted to many others whose ideas and works were collected for use in this book. Special thanks are given to the following people: the author's supervisor, Dr. Anthony Wachinski, who encouraged him to write the book. Without his enormous support and contributions, the book would not be a reality. Mr. Brent Alspach of Malcolm Pirnie and Mr. James Vickers of the SPI reviewed the manuscript and provided constructive critiques and many insights. Many current and former colleagues in the Pall Corporation have contributed to the book directly or indirectly. Mr. James Schafer exchanged many ideas with the author on this subject and helped to articulate many concepts. Mr. James Moy invented the fiber cutting tool that is instrumental to the fiber-cutting demonstration. He also helped perform investigations of assessing membrane integrity. Dr. Amarnauth Singh provided understanding of the membrane manufacturing process and provided electronic microscope images. In addition, Mr. Takehiko Otoyo of Asahi Kasei Chemicals Corporation provided many valuable suggestions.

The author would like to express his gratitude to Dr. Sandeep Sethi of Carollo Engineers and the members of the research team for AwwaRF project *Assessment and Development of Low-Pressure Membrane Integrity Monitoring Tools*, for which the author served as a member of project advisory committee, and to Dr. Jarka Povovicova and Ms. Tracy Case, the AwwaRF project managers. Much of chapter 6 is based on the data generated in the project.

Thanks to AwwaRF (now known as the Water Research Foundation), the American Chemical Society, and Dr. Karl Bruce Jacobson of Oak Ridge National Laboratory for permitting the use of copyrighted materials. Thanks to Drs. Jirachote Pattaranawik, A.G. Fane, and F.S. Wong of the Nanyang Technological University for sharing their data on relative trans-membrane pressure (TMP) monitoring techniques; Michael Sadar for sharing the data on multiplex integrity monitoring; Dave Thomas of Brightwell Technologies for providing the photograph of the Micro-Flow Imaging System; Kent Peterson of Fluid Imaging Technologies for providing the photograph of the FlowCam® System; and Dr. Amarnauth Singh of Pall Corporation and Takehiko Otoyo of Asahi Kasei Chemical Company for providing SEM photographs of membranes.

Linda Bevard, the manuscript editor, did a wonderful job revising the manuscript. Her patience and tireless efforts caught mistakes in the manuscript and definitely made the book much more readable. Gay Porter De Nileon, the publications manager of AWWA, managed the process.

Lastly, the author would like to express his gratitude to his wife, Jianying, and his children, Hervey and Helen. Their encouragement and good-natured teasing got the author across the finish line.

1

Membrane Integrity Testing in the Drinking Water Industry: A Historical Perspective

COMING OF AGE: LOW-PRESSURE MEMBRANES IN DRINKING WATER FILTRATION

Low-pressure membrane filtration refers to microfiltration (MF) and ultrafiltration (UF). The technology has become increasingly popular in the drinking water industry since the mid-1990s. Its popularity is the result of the combination of several factors.

The primary driver of this technology in the drinking water industry is regulations for drinking water quality. Historically, the evolution of water treatment technologies is relatively slow. The first water treatment plant serving a municipal water supply including a settling basin, and a gravel and a sand filter was installed in Scotland in 1804. Chlorine in the form of bleach powder was first used to purify foul-smelling water in England in 1830s. In the United States, the late 1800s and early 1900s were the golden years of progress in water treatment. Slow sand and rapid sand filters were introduced in 1872 and 1882, respectively. Chlorination was introduced around 1910. (For a complete account of the early history of water treatment, see Symons 2006.)

Since the early 1900s, although incremental improvements and sometimes significant innovations occurred from time to time, the practices for treating municipal water supplies in the United States seemed by and large to be kept in a time capsule. For surface water sources, treatment processes consisted of coagulation, settling, and filtration with granular media such as sand, gravel, crushed anthracite coal, and granular activated carbon (GAC); and disinfection primarily with chlorine became the gold standard. These practices started to change mainly due to two significant events that occurred about two decades apart. The first event was the discovery of trihalomethanes in the mid-1970s and, later, of other disinfection by-products (DBPs). These discoveries shook the foundations of water treatment practice and led to new regulations for disinfection practices.

The second event was the discovery of chlorine-resistant pathogens in the late 1980s and early 1990s. The outbreak of cryptosporidiosis in Milwaukee, Wisconsin, in 1993 brought the quality of drinking water to public attention. The widely reported event was blamed for having infected over 400,000 people and causing more than 100 deaths. This discovery, in conjunction with the issues of DBPs, posed a Catch-22 for many water suppliers. On the one hand, disinfection must be carried out cautiously to minimize the generation of DBPs, which often requires a reduction of either the chlorine dose or the contact time or of both; on the other hand, the reduction of chlorine-resistant pathogens requires increased chlorine dose and/or time for chlorine exposure.

As a result, alternative disinfectants and/or treatment processes that could meet both treatment goals suddenly become more attractive. The first low-pressure membrane plant in the world was installed in 1984, while the first low-pressure membrane plant in the North America was installed in 1989 (Adham et al. 2005). The number of low-pressure membrane plants and their cumulative capacity in million gallons per day (mgd) in North America in the past decade are plotted in Figure 1-1. Important regulatory milestones that might have driven the low-pressure membrane market in the United States are also noted on the figure.

The data in Figure 1-1 were dated to 2003. Since then, the pace of growth of the membrane market in the United States and around the world has been fast. Two new regulations in the United States, the Long-Term 2 Enhanced Surface Water Treatment Rule (LT2ESWTR) and the Stage 2 Disinfectants/Disinfection By-Products Rule (Stage 2 D/DBP), were released in 2006 (USEPA 2006a; USEPA 2006b) and will continue to fuel the growth of the membrane market in the United States for years to come.

Besides increasingly stringent regulations, a water shortage is also driving the adoption of membrane technologies in the industry. Whether as a result of population growth or drought, regions not previously considered to have water shortages do now. This shortage forced utilities to use sources that are low in quality, such as wastewater, secondary effluent, brackish water, and seawater. To treat water from these sources to acceptable quality is often beyond the realm of conventional water treatment processes. However, low-pressure membranes are advantageous in treating poor quality water because unlike conventional treatment processes, the filtrate quality from membrane processes is not affected by the quality of raw water. This also makes low-pressure membranes excellent choices as pretreatment of advanced processes such as nanofiltration (NF) and reverse osmosis (RO), which are used to the remove dissolved contaminants and to desalinate brackish and seawater.

MEMBRANE INTEGRITY TESTING IN DRINKING WATER—A BRIEF HISTORY

The typical pore size of low-pressure membranes used for drinking water filtration is 0.2 μm (1 μm is one-millionth of a meter) or less. This would enable the membranes to remove nearly all microbial pathogens except viruses, as long as the integrity of the membranes remained intact. Integrity refers to wholeness, being unimpaired

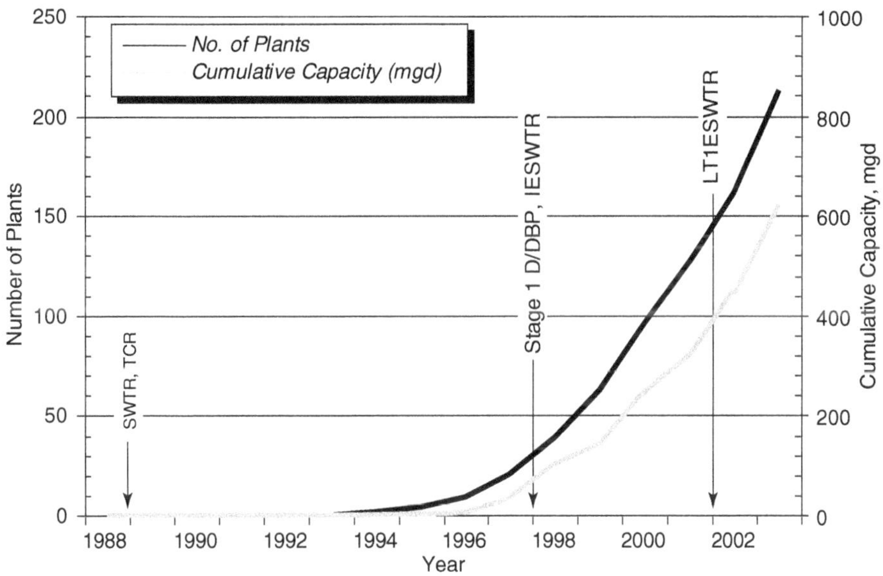

Figure 1-1. The number of low-pressure membrane plants and cumulative capacity in million gallons per day in North America: 1998–2003 (plotted from data of Adham et al. 2005)

or free of defects. Integrity testing uses testing methods to identify and locate the defects. Membranes, such as MF and UF, have a long history of application in the pharmaceutical industry, where the removal of contaminants from products is of paramount importance. Integrity testing has its roots in these applications, especially when filter sterilization is required.

In the United States, the Food and Drug Administration (USFDA) established the requirements for the integrity testing of filters (FDA 1976). Most of the applications require testing the integrity of filters in-situ. Therefore, testing has to be nondestructive. The removal efficiency of the targeted contaminant must be validated and verified through integrity testing.

Various methods for membrane integrity testing have been developed. A detailed account of integrity tests, including bubble-point test, pressure-hold test, and diffusive air-flow test for filters used in pharmaceutical industry, is provided by Meltzer (1987). One of those tests, known as *bubble-point test*, was codified into a Health Industry Manufacturers Association (HIMA) standard and later into the American Society of Testing and Materials (ASTM) methods (Moch and Paulson 2003).

The rapid spread of low-pressure membranes in the drinking water industry raised the need to establish testing standards of membrane integrity verification (Adham, Jacangelo, and Laine 1995). The United States Environmental Protection Agency (USEPA) published a technical document titled *Low-Pressure Membrane Filtration for Pathogen Removal: Application, Implementation, and Regulatory Issues* (USEPA 2001). The issues of membrane integrity testing were discussed extensively and many concepts presented in the document were utilized and expanded in the *Membrane Filtration Guidance Manual* published by USEPA four years later (USEPA 2005). At the same time, standard-setting and research activities started to flourish in the drinking water industry.

In retrospect, three significant events occurred during the latter 1990s and early 2000s. These events may have had important influences on membrane integrity testing in the drinking water industry in the United States.

- ASTM Standard Practice for Integrity Testing of Water Filtration Membrane Systems, ASTM D-6908-03 (ASTM 2003b)
- The research project of American Water Works Association Research Foundation on assessment and development of low-pressure membrane integrity monitoring tools
- The regulatory framework crafted in the Long-Term 2 Enhanced Surface Water Treatment Rule and *Membrane Filtration Guidance Manual* (MFGM)

ASTM Standard Practice for Integrity Testing of Water Filtration Membrane Systems (ASTM D-6908-03)

ASTM formed a subcommittee, D19.08.02, to undertake the task of writing a membrane integrity standard in 1997. Recognizing that many integrity testing methods were proprietary and product-specific, the subcommittee attempted to standardize and unify various practices of integrity testing used by different parties and to publish them as standard practices for the drinking water industry. ASTM standards were developed by members' consensus. The subcommittee had a diversified membership including manufacturers, consultants, regulatory agencies, and end-users.

To manage the number of the competing methods of integrity testing, the subcommittee used an advocate approach—members volunteered to write the draft describing the selected methods and to defend their validity. After much debate, revision, updating, and data compilation, the standard practices—designated as ASTM D-6908-03—finally went through the balloting process and were published in June 2003. Moch and Paulson (2003) provided a detailed historical account of the developments of the standard practices.

ASTM D-6908-03 includes three practices:

1. Pressure-decay test (PDT) and vacuum-decay test (VDT)
2. Use of total organic carbon (TOC) analyzers for monitoring integrity of reverse osmosis and nanofiltration membrane systems
3. Soluble dye test

Among the three practices, only Practice 1 is feasible for low-pressure membranes (although it can also be applied to NF and RO for fairly large integrity breach in the system). Practices 2 and 3 are for NF or RO.

The significant aspects of ASTM development of the Standard Practices include:

- It provided a rather detailed description of each practice used in membrane integrity testing.
- It unified the PDT and VDT with different testing patterns (i.e., whether the high pressure is on the filtrate side or feed side of membranes).
- It presented the equation relating testing pressure and the size of defect.
- It presented the method relating the removal efficiency expressed as log reduction value (LRV) to the system conditions.
- It presented the concept of using equivalent number of broken fibers for correlating testing results of pressure decay test to LRV.

The ASTM Standard Practices also have some limitations. For example, the Practices do not address the issue of sensitivity and reliability of each practice, thereby failing to inform a potential user about the practical limitations of a particular practice. In addition, Standard Practice A uses the Hagen-Poiseuille Equation to calculate both water and air flow through a cylindrical defect under the assumption that both air and water bypass flow are laminar, which limits its usage to those conditions.

AwwaRF Research Project on Integrity Monitoring of Low-Pressure Membranes (RFP 2861)

On a parallel path to the development of ASTM Standard Practices, American Water Works Association Research Foundation (AwwaRF) (currently the Water Research Foundation, WRF) funded a project titled *Assessment and Development of Low-Pressure Membrane Integrity Monitoring Tools* (RFP 2681, AwwaRF 2000). During the development of the request for proposal (RFP), several performance criteria for assessing a method for integrity-testing were suggested to be added to the proposal (Povovicova 2000). These criteria include:

- *Sensitivity:* To what degree a membrane breach can be detected with a given integrity testing procedure?
- *Reliability:* How reliable is the result of an integrity testing procedure? What is the probability of a false positive or a false negative?
- *Identifiability:* Is the integrity testing procedure able not only to *detect* but also to *locate* the defect? If so, how soon?
- *Continuity:* How often can the system integrity be monitored? Is the frequency of the integrity testing sufficient to warrantee the regulatory compliance?
- *Economy:* Are the additional costs associated with integrity testing, capital, and operations and maintenance (O&M) reasonable and affordable for the end-users?

Coincidentally, the RFP addressed most of the same issues addressed in the LT2ESWTR. The three specific objectives of the research project were the following (Sethi et al. 2004):

- Identify and document existing integrity monitoring methods through literature review and survey of MF/UF plants
- Perform testing of existing and new integrity monitoring methods through full-scale testing at facilities representing the range of commercially available MF and UF systems
- Assess integrity monitoring methods based on the results of the full-scale testing and the five essential criteria—sensitivity, continuity, identifiability, reliability, and implementability

The key findings and conclusions of the research project are (Sethi et al. 2004):

- Sensitivity and reliability criteria should be considered uniquely important regardless of other criteria, such as continuity, identifiability, and economy, because the lack of either in practical applications can be a fatal flaw.
- Among the tested methods for integrity monitoring, the pressure decay test, microbial challenge test (with *Bacillus subtilis*), and high sensitivity particle counters (0.05 or 0.5 μm) were found and/or inferred to be the most sensitive and reliable methods.
- Indirect methods (i.e., methods based on monitoring filtrate quality) showed sensitivity and reliability that are affected by the differences in systems (e.g., *with* versus *without* air-assisted backwash) and feed water quality.
- For indirect methods, reliability is an important issue in terms of practical application especially for eliminating false positives and false negatives. Even though an indirect method might be highly sensitive, its successful application may require adequately detailed statistical analysis to achieve sensitivity and reliability.

The AwwaRF research project was the first to provide the most comprehensive and scientifically sound data at the production plant scale. Moreover, the project interpreted the data in the context of issues that are critical to membrane integrity testing in the drinking water industry. One limitation of the project is the lack of correlating LRV and testing parameters in pressure decay tests, although an equation to calculate LRV based on hydraulic modeling was developed and the data for validating the correlation were available. Nevertheless, the project is a significant milestone in the practice of membrane integrity monitoring in the drinking water industry.

LT2ESWTR and MFGM

Other watershed events in the evolution of concepts and practices of membrane integrity testing in the drinking water industry were the publication in 2005 of the *Membrane Filtration Guidance Manual* (USEPA 2005) and the promulgation in 2006 of the LT2ESWTR (USEPA 2006a).

Before those two events, there were significant differences among state agencies in terms of the requirements for granting removal credit and integrity testing. According to a survey (USEPA 2001), 14 of 29 states that had installed membrane systems did not require any integrity monitoring for membrane plants aside from turbidity. Of the remaining 15 states, 7 required both continuous indirect monitoring and periodic direct integrity testing, 5 required only direct integrity testing, 1 required only continuous indirect monitoring, and 2 approved of integrity monitoring plans on a case-by-case basis (USEPA 2001). The report explained the reasons for the situation.

> Much of the variability in state requirements for membrane filtration is a result of factors such as different approaches to multiple barrier treatment, different levels of experience with membrane processes, the lack of standardization in the technology field, and a lack of formal guidance from USEPA that adequately addresses this technology. This variability presents challenges for the implementation of this technology.

To address these issues, the USEPA drafted guidelines for granting removal credit and requirements for integrity testing, anticipating the increasing demand for membrane technology in the drinking water industry driven by the impending LT2ESWTR.

The purpose of the LT2ESWTR is very clear and specific—to manage the microbial risk posed by disinfectant-resistant *Cryptosporidium*. However, unlike previous drinking water regulations, the LT2ESWTR takes a different approach. First, it classifies the risks of *Cryptosporidium* in drinking water into four bins based on the monitoring results of the source waters. The bin designation determines what additional level of treatment for *Cryptosporidium* is required beyond conventional treatment. To assist utilities in complying with the rule, USEPA includes a microbial toolbox—the options for compliance with the rule and the removal/inactivation credits associated with each option. Table 1-1 presents the microbial toolbox.

As indicated in Table 1-1, membrane processes are included in the toolbox. However, unlike all other tools enlisted, there are no granted credits for membrane and ultraviolet (UV). Rather, the removal credits for membrane processes are only granted based on "removal efficiency demonstrated in a challenge test for a device if supported by direct integrity testing" (USEPA 2006a). The LT2ESWTR uses a framework regulating membrane processes for rule compliance consisting of the following major provisions:

Table 1-1. Microbial toolbox for compliance with the LT2ESWTR

Toolbox Option	***Cryptosporidium* Treatment Credit With Design and Operational Criteria***
Source Protection and Management Toolbox Options	
Watershed control program	0.5-log credit for state-approved program comprising required elements, annual program status report to state, and regular watershed survey. Unfiltered public water systems (PWSs) are not eligible for credit.
Alternative source/intake management	No prescribed credit. PWSs may conduct simultaneous monitoring for treatment bin classification at alternative intake locations or under alternative intake management strategies.
	Prefiltration Toolbox Options
Presedimentation basin with coagulation	0.5-log credit during any month that presedimentation basins achieve a monthly mean reduction of 0.5-log or greater in turbidity or alternative state-approved performance criteria. To be eligible, basins must be operation continuously with coagulant addition and all plant flow must pass through basins.
Two-stage lime softening	0.5-log credit for two-stage softening where chemical addition and hardness preciptation occur in both stages. All plant flow must pass through both stages. Single-stage softening is credited as equivalent to conventional treatment.
Bank filtration	0.5-log credit for 25-ft setback; 1.0-log credit for 50-ft setback; horizontal and vertical wells only; aquifer must be unconsolidated sand containing at least 10 percent fines (as defined in rule); average turbidity in wells must be less than 1 ntu. PWSs using existing wells followed by filtraiton must monitor the well effluent to determine bin classification and are not eligible for additional credit.
Treatment Performance Toolbox Options	
Combined filter performance	0.5-log credit for combined filter effluent turbidity less than or equal to 0.15 ntu in at least 95 percent of measurements each month.
Individual filter performance	0.5-log credit (in addition to 0.5-log combined filter performance credit) if individual filter effluent turbidity is less than or equal to 0.15 ntu in at least 95 percent of samples each month in each filter and is never great than 0.3 ntu in two consecutive measurements in any filter.
Demonstration of performance	Credit awarded to unit process or treatment train based on a demonstration to the state with a state-approved protocol.

(Table continued next page)

Table 1-1. Microbial toolbox for compliance with the LT2ESWTR (continued)

Toolbox Option	***Cryptosporidium*** **Treatment Credit With Design and Operational Criteria***
	Additional Filtration Toolbox Options
Bag and cartridge filters	Up to 2-log credit with demonstration of at least 1-log greater removal in a challenge test when used singly. Up to 2.5-log credit with demonstration of at least 0.5-log greater removal in a challenge test when used in series.
Membrane filtration	Log credit equivalent to removal efficiency demonstrated in challenge test for device if supported by direct integrity testing.
Second-stage filtration	0.5-log credit for second separate granular media filtration stage if treatment train included coagulation prior to first filter.
Slow sand filters	2.5-log credit as a secondary filtration step; 3.0-log credit as a primary filtration process. No prior chlorination.
	Inactivation Toolbox Options
Chlorine dioxide	Log credit based on measured CT in relation to CT table.
Ozone	Log credit based on measured CT in relation to CT table.
Ultraviolet (UV)	Log credit based on validated UV dose in relation to UV dose table; reactor validation testing required to establish UV dose and associated operation conditions.

* Table provides summary information only.
Source: Adapted from USEPA 2006a.

- All membrane systems used for compliance with the rule must undergo periodic direct integrity testing and continuous indirect integrity monitoring. (The discussion of direct versus indirect integrity testing will be detailed in chapter 4).
- The removal credit granted for a membrane system is based on the lower of (a) the result of a product-specific challenge test and (b) the result of a system-specific integrity testing.
- The granted removal credit cannot exceed what can be verified through integrity testing.
- (Direct) integrity testing must meet certain performance criteria.
- For integrity testing to be applied, an upper control limit (UCL) must be established to signal the occurrence of an integrity breach.

USEPA published *Membrane Filtration Guidance Manual* to assist utilities that intend to use membranes as the tools for rule compliance. In this manual, the issues of granting removal credits and membrane integrity testing are discussed extensively.

The LT2ESWTR and the guidance manual took an open-ended and accommodating approach to future technological development in membrane integrity testing. That is, specific integrity testing methods were not stipulated in the requirement. Rather, no matter what the methods are, they must meet certain performance standards consisting of three specific criteria—resolution, sensitivity, and testing frequency. A detailed discussion of the requirements and implications of the LT2ESWTR and MFGM is presented in chapters 2 and 3.

Because USEPA is the most authoritative source on regulatory matters, the part of the LT2ESWTR concerning membrane filtration and the MFGM undoubtedly greatly influence practices in the drinking water industry. They have quickly become the model template used by state regulatory agencies as they draft their own regulations on membrane processes. Moreover, the LT2ESWTR's regulatory framework has influence beyond compliance and permeates all regulations applying to situations where membrane filtration is used as alternative filtration technology (AFT). This intention is clearly stated in the introduction of the MFGM.

> This regulatory framework and associated guidance manual may also be useful resources for states to develop distinct regulatory policies for membrane filtration relative to existing surface water treatment rules (i.e., those promulgated prior to the LT2ESWTR) to separate the technology from the broader AFT classification. (USEPA 2005)

On the technical issues, the MFGM addresses the limitation on flow regime that the ASTM standard has by adding equations for LRV calculation for different membrane configurations (hollow fiber and flat sheet) and different flow regimes (laminar and turbulent). Therefore, it expands the usefulness significantly. However, the assumption that water bypass flow and air flow through a membrane breach are both in the flow regime may not be applicable for all the circumstances. The MFGM does not to address the issues of reliability of integrity testing, even though it has been one of the outstanding issues raised by several states (USEPA 2001). In addition, the MFGM may have attempted to cover too many integrity testing methods, whereas the pressure-decay test (PDT) is performed by 98 percent of low-pressure membrane plants in use (Vickers 2009). Finally, even though in principle PDT can be used for NF/RO plant, it is not really applicable on a commercial level (Vickers 2009). In spite of those shortcomings, the LT2ESWTR and the MFGM are pivotal in the evolution of concepts and practices of membrane integrity testing and will continue to have profound impacts on membrane processes used in the drinking water industry.

By and large, the technologies used for integrity testing and monitoring have been available for decades. As sensors, data-processing, and communication technologies evolve, new concepts and methods for membrane integrity testing will emerge. It is conceivable that soon after this book is printed, new concepts and technologies for membrane integrity testing may emerge—making this subject dynamic and exciting.

2

Background: What Is Membrane Integrity Testing and Why Do It?

LOW-PRESSURE MEMBRANES AND MEMBRANE SYSTEMS—A PRIMER

Before discussing membrane integrity, it is beneficial to introduce the concepts of membranes, membrane filter modules, and skid. A basic knowledge of how membranes are constructed, configured, and assembled into a working system will be helpful to understand membrane integrity issues.

Low-Pressure Membranes

Low-pressure membranes include microfiltration (MF) and ultrafiltration (UF) membranes. The separation mechanism for both MF and UF is based on size exclusion, or mechanical sieving. That is, anything that is larger than the pore size would be physically retained by the membrane. This interpretation may be simplistic because of the following reasons:

1. The pore size of a membrane is not always absolute as pores are not all the same in size and shape. Therefore, pore size of a membrane may be a statistical concept.
2. Particles smaller than membrane pores could land on the edge of a pore so they could be still retained by the membrane.
3. Microscopically, many membrane pores are actually tortuous pathways so that particles smaller than pore size can still be retained within those pathways.
4. Accumulation of previously retained particles can form a barrier with effectively smaller pores.
5. Local hydrodynamic conditions can prevent particles with a certain size range from reaching the membrane, even though the particles may be smaller than pore sizes.
6. Mechanisms other than mechanical sieving can retain particles smaller than pore size of membranes. One example is retaining natural organic matter on membrane surface by membrane–particle interactions rather than mechanical sieving.

7. The retention of nonspherical particles may depend on the orientation of the particle relative to the pores.
8. Conceptually, a deformable particle (e.g., an air bubble or a liquid droplet) can pass a pore smaller than its size by deformation under stress.

Nevertheless, if the maximum pore size of a given membrane is known, any nondeformable particles larger than this size can surely be retained by the membrane. For particles that are nondeformable and smaller than the maximum pore size of the membrane, there could be a problem in distribution for their removal, depending on many factors.

The distinction of MF and UF is not exact, and there is no definitive boundary on where pore-size microfiltration ends and ultrafiltration begins. To complicate the matter further, MF is rated by pore size of the membrane in micrometers whereas some UF membranes are rated by a unit termed *molecular weight cutoff* (MWCO) expressed as daltons, while other UF membranes are rated by pore size (AWWA 2005). The pores of UF membranes are small—about one order of magnitude or more smaller than those of MF membranes. With MF, the pores on the membrane can be directly visualized and measured under the scanning electronic microscope (SEM), but this cannot be done on the pores of the UF membranes.

The terms *microfiltration* and *ultrafiltration* were crafted long before the invention of the atomic force microscope (AFM), which has a resolution of a fraction of one Angstrom (one tenth of a nanometer). The way to rate a UF membrane is typically by its rejection property obtained through a challenge test. In a challenge test, marker reagents with known molecular weight and concentration are injected in the feed, and the concentration of the markers in the filtrate is measured. The molecular weight of the marker corresponding to a certain rejection (e.g., 90 percent) is rated as nominal molecular weight cutoff.

Similarly, the molecular weight of the marker corresponding to a 100 percent rejection is rated as absolute molecular weight cutoff. Because the passage of a marker through an UF membrane pores depends not only its molecular weight, but also the shape of the molecules, as well as the distribution of pore sizes, there is no definitive correlation between an UF rating based on MWCO and pore size. The general consensus is that UF have pore sizes typically below 0.05 μm.

In the context of contaminant removal, MF is used to remove particles (0.1 μm or larger), pathogenic protozoa such as *Cryptosporidium* and *Giardia*, and bacteria, whereas UF is used to remove all of the above plus viruses. For the purpose of meeting treatment goals, rating membranes by their ability to retain a target organism via microbial challenge test instead of pore size is more meaningful. The use of an atomic force microscope enables the direct measurement of pore size of UF. Therefore, the confusion in rating MF and UF membranes can be finally clarified with the AFM technique.

Figure 2-1. Symmetric (*left*) versus asymmetric membrane (*right*)
(Courtesy of Dr. A. Singh of Pall Corporation and Mr. T. Otoyo of Asahi Kasei Chemical Company)

If the morphology of pore shape along the thickness of a membrane is observed, membranes can be classified as *symmetric* and *asymmetric.* An asymmetric membrane has a tapered pore in which the pores on one side of the membrane are larger than on the other. Figure 2-1 illustrates the difference of symmetric and asymmetric membranes. The symmetric membrane (left in Figure 2-1) shows a consistent density of membrane medium across the section, while the asymmetric membrane shows significant changes in the medium density. Sometimes, the terms *isotropic* and *anisotropic* are also used to describe symmetric and asymmetric membranes, respectively. The idea of making asymmetric membranes is to reduce hydraulic resistance along the flow path in the membrane medium while maintaining its size-exclusion characteristics. In this case, the side of the membrane with smaller pores, typically facing the feedwater, defines the rating of the membrane.

Asymmetric membranes are made by either phase inversion or interfacial polymerization—coating a layer of dense medium over a layer of medium with higher voids and loose "pore size." These processes result in a membrane with a dense layer as the functioning membrane layer to define retention characteristics and a loose layer called *substrate* to provide physical support. Sometimes, the membrane layer and supporting layer are made of two different media. Therefore, we may also classify the membranes into mono-cloth and dual- or multi-cloth membranes according to the composition of membrane media.

The composite membrane is mostly used for reverse osmosis (RO) and nanofiltration (NF) membranes. It is also used for some UF membranes, as depicted in Figure 2-2.

Most of the low-pressure membranes used for drinking water filtration are made of polymeric materials, although recently interest in other materials such as ceramic is rising in some parts of the world (Clements et al. 2006). Table 2-1 lists the most

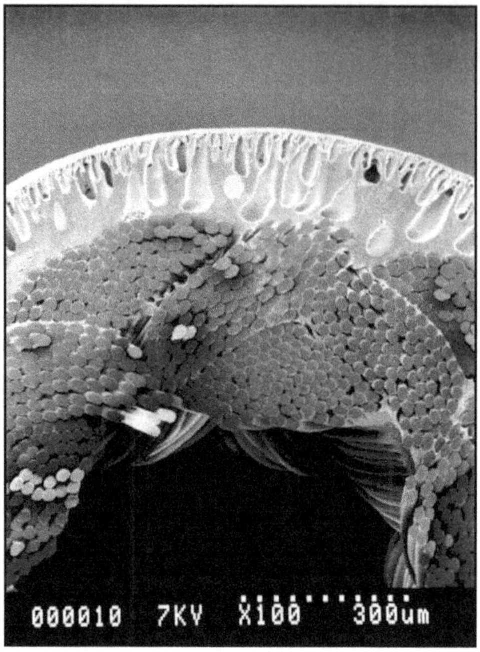

Figure 2-2. A dual-cloth composite membrane

common polymeric materials used for manufacturing membranes. According to a survey by Awwa Research Foundation (AwwaRF) in 2005, over one half of the installed capacity in million gallons per day (mgd) of different MF/UF membrane materials is polyvinylidene fluoride (PVDF). Other membrane materials—polyethersulfone (PES), polysulfone (PS), polypropylene (PP), and cellulose acetate (CA)—range from 8 percent to 15 percent of the installed capacity (Adham et al. 2005). The dominance of PVDF membranes may relate to the general characteristics of high chemical resistance and mechanical strength.

Most polymeric membranes for drinking water applications are in hollow fiber configuration. Other configurations of low-pressure membranes are also available but are less popular in drinking water applications. Therefore, this discussion will focus on polymeric hollow-fiber membranes. A hollow fiber is a long, fine "tube" with permeable walls consisting of membranes. The inner diameter of hollow fibers may range from 0.2 mm to ~1 mm. The thickness of the walls is typically in the range of 0.1 to 0.4 mm. The typical length of the tube can vary from 1 m to 2 m.

The hollow-fiber configuration has two distinctive advantages over other configurations.

1. The configuration provides a structural stability that enables the membranes to sustain mechanical stresses from operations (e.g., backwash and air scoring).
2. The configuration provides the possibility of packing membranes densely as to reduce the footprint.

Table 2-1. Common polymeric materials for microfiltration and ultrafiltration membranes

Polymer	Chemical Structure	T_G (°C)	T_M (°C)
Cellulose acetate (CA)	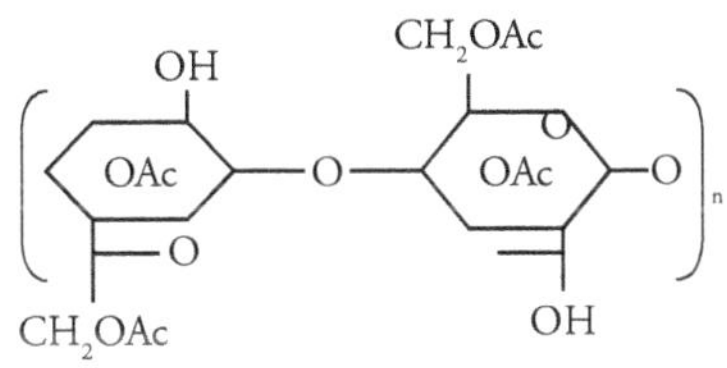	67 to 68	150 to 230
Polyacrylonitrile (PAN)	$(CH_2\text{-}CH(CN))_n$	80 to 110	320
Polyethersulfone (PES)	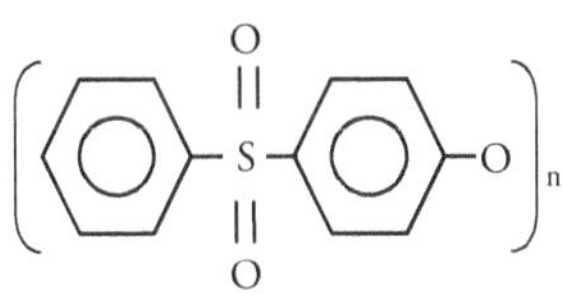	225	N/A
Polyethylene (PE)	$(CH_2)_n$	–90 to –30	137 to 145
Polysulfone (PS)	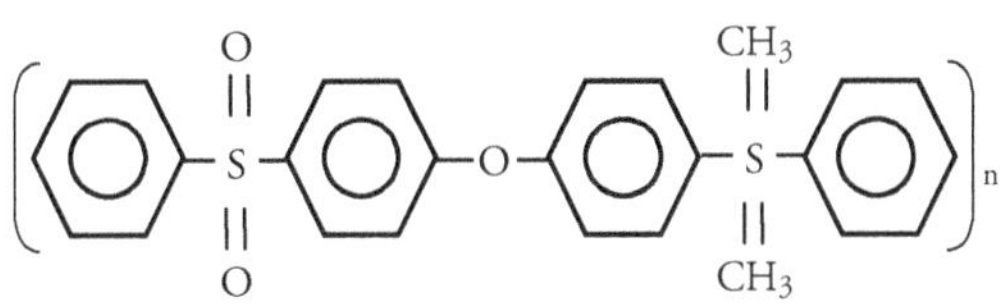	190 to 250	N/A
Polyvinylalcohol (PVOH)	$(CH_2\text{-}CH(OH))_n$	65 to 85	230 to 260
Polyvinylidene fluoride (PVDF)	$(CH_2\ CF_2)_n$	–50 to –35	160 to 185

T_G = glass transition temperature
T_M = melting point
N/A = not available

Hollow-fiber membranes may be used in two modes: *inside out* or *outside-in*. In inside-out mode, feedwater flows inside the tube (also called *lumen*) and the filtered water comes out the semi-permeable walls. In outside-in mode, feedwater flows outside, through the membrane walls, and into the lumen.

Membrane Filter Module and Membrane System

A membrane filter module is the smallest modular unit manufactured as a product. To assemble hollow-fiber membranes into a filter module, a bundle of hollow fibers (from a few hundreds up to tens of thousands) are potted with resins to isolate the feedwater and filtrate water. A filter module may have a shell or housing to house the hollow-fiber

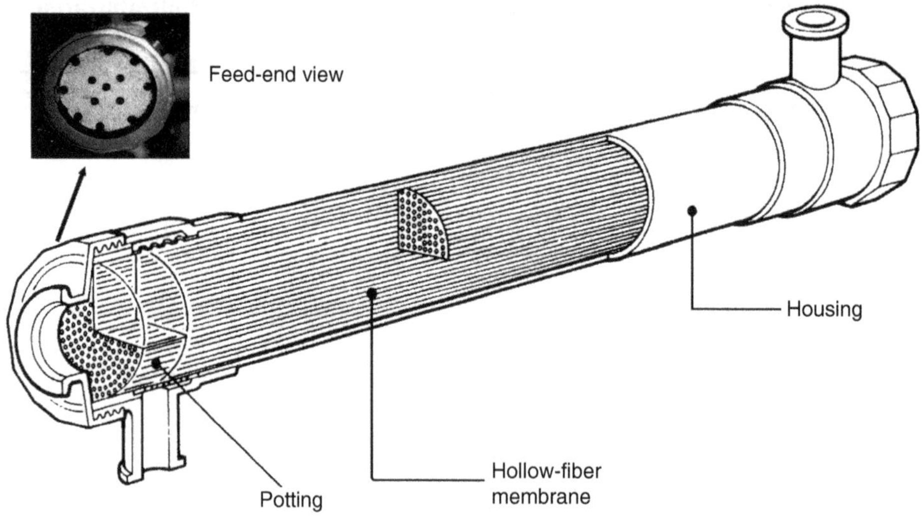

Figure 2-3. Schematic of a shelled hollow-fiber membrane module

bundle and to regulate flow. Alternatively, the fiber bundle may not have a shell but may have resin potting to keep feedwater and filtrate separated. Each module has ports for feedwater, filtrate, and sometimes a circulation loop. Figure 2-3 presents a schematic diagram of a shelled hollow-fiber membrane module.

A membrane system may consist of one or multiple filter modules, hardware, control and instrumentation, and so on, to make it functional. For large drinking water plants, there may be hundreds or up to tens of thousands of modules. Those filter modules may be divided into groups (skids) that are connected by a piping network. Depending on the size of a system, its output can range from tens of thousands of gallons to over one hundred million gallons per day (gpd). Figures 2-4 and 2-5 show various membrane systems. One feature of membrane system is its modularity, which makes the scale-up of a system relatively easy. One can add additional filter modules on a skid and increase the number of skids to increase outputs.

MEMBRANE INTEGRITY, INTEGRITY BREACH, AND INTEGRITY TESTING

Membrane Integrity

According to the *American Heritage Dictionary*, the term *integrity* refers to (a) the state of being unimpaired, soundness, and (b) the quality or condition of being whole or undivided, completeness. When the integrity of membranes is discussed, the team refers to the state or conditions of being unimpaired or defect-free. Low-pressure membranes are used as the barriers for the microbial pathogens and particles in water. Therefore, an

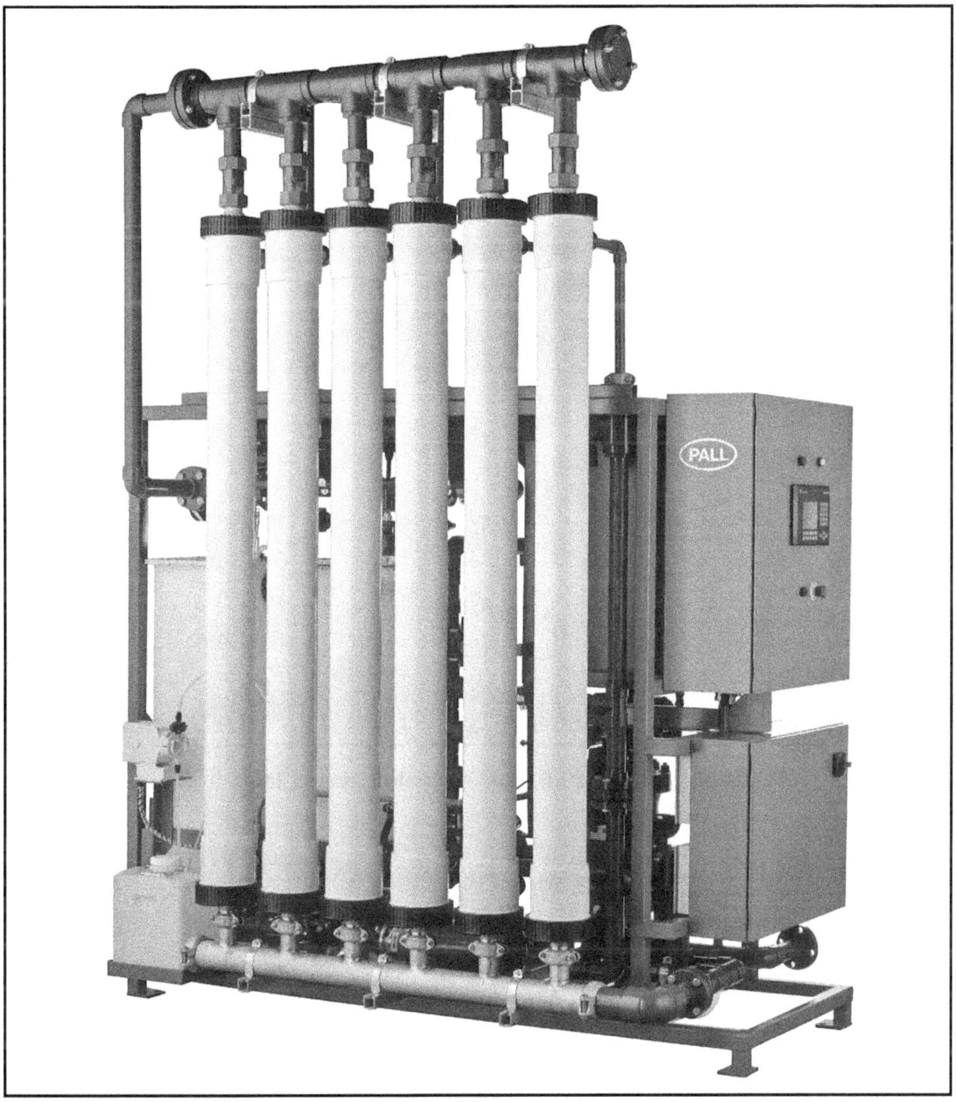

Figure 2-4. Microfiltration plant with capacity of 70,000 gpd

integral membrane would prevent any of the microbial pathogens and particles with a size above its size of exclusion from entering into the filtrate stream under the designed operating conditions. If the integrity of a membrane is compromised (i.e., the membrane has a defect or a breach), there is a chance that pathogens would pass through the defect and enter into the filtrate stream.

The flow through the defect is termed *bypass flow*. Therefore, another way to define membrane integrity is as an integral system free of any bypass flow. The impact of an integrity breach on filtrate water quality is a function of the ratio of bypass to total filtrate flow of a given system. Integrity testing is a procedure or a set of procedures used to

Figure 2-5. Membrane skids in 20-mgd microfiltration plant

determine (a) whether a membrane system is integral, (b) if not, how large the integrity breach is, and (c) where the breach is located.

Types of Integrity Breach

There are different types of integrity breaches for a hollow-fiber membrane module, ranging from *pinholes* in the membranes to partially or completely severed fibers. A pinhole is tiny hole that may or may not be visible to the naked eye as the resolution of the naked eye of a human being is not greater than 50 μm. However, it is certainly large enough to allow pathogens originally retainable by the same intact membrane to pass into the filtrate. Pinholes are most likely created during the manufacturing process of membranes.

For composite membranes, *delaminating* can also occur. Delaminating refers to the phenomenon that in a multi-cloth membrane, the top membrane layer is peeled from the underlying supporting layer. This would result in the loss of integrity as the supporting layer does not have the same rejection as the membrane layer.

Some modules employ O-rings in assembling to isolate filtrate from feedwater. Occasionally, an O-ring can also be pinched or incorrectly alimented during assembling. This would also create opportunities for bypass.

CAUSES OF INTEGRITY BREACH

There are several sources contributing to membrane integrity breaches. Namely,

1. Manufacturing defects
2. Improper design, installation, and operation and maintenance
3. Long-term wear and tear

The following sections will discuss each cause separately.

Manufacturing Defects

Manufacturing hollow-fiber membrane modules involves the production of hollow fibers and assembling fibers into modules. Hollow fibers are prepared by spinning a solution of base material into a coagulation bath. Phase inversion can be induced by either chemical or thermal means. In the case of composite membrane, the supporting medium is spun into hollow fiber first, and the membrane layer is coated through interfacial polymerization, in which supporting medium is immersed in an aqueous solution of monomer and then in a water-immiscible solution of the second monomer to form a very thin membrane layer. After hollow fibers are made, they are cut and potted into module. Potting of hollow fibers is another important step. The failure of potting is also a source of integrity breach because a potting layer separates upstream and downstream of membranes.

The manufacturing of hollow-fiber membrane filters is a multiple-step process, and many factors can contribute to defects. For example, impurities or air bubbles, if they remain in the base-material solutions or casting solutions for membranes, can cause pinholes in membranes. Another factor is the effect of the process control (e.g., the cooling rate) on polymer crystallization (Kumar and Gupta 1998), which directly relates to a membrane's mechanical strength and chemical stability. The key is to establish good quality assurance and quality control (QA/QC) programs to prevent the defects from occurring in the first place.

Improper Design, Installation, and Operation and Maintenance

The polymeric membranes are of certain mechanical strength and chemical stability. Beyond those limits, membranes can deteriorate or be damaged. Therefore, the proper design, installation, and operation and maintenance of a membrane system are critical to maintain its integrity. It is important to understand that integrity depends on not only membranes but also the overall system. Design and installation should consider avoiding water hammers during the opening and closing valves and reducing the vibration of membrane skid—both can cause the cracks in filter housing and piping and add mechanical stress on membranes.

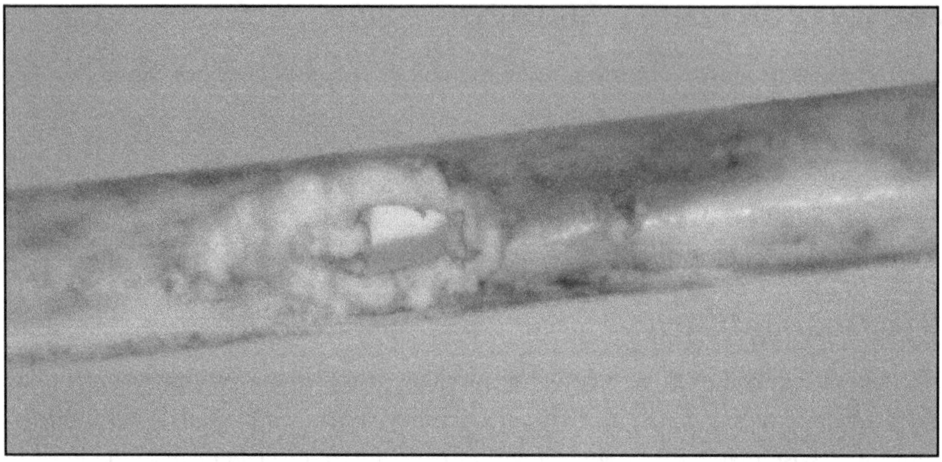

Figure 2-6. Hollow-fiber membrane damaged by construction debris

One example of improper installation is insufficient flushing of debris introduced during the construction in the system prior to startup. The remaining debris in the system can damage membranes once the system is operational. Figure 2-6 shows a hollow-fiber membrane damaged by the debris in the piping. The rough perimeter indicates cutting by a sharp-edged object, which was also found in the same module.

One operation integrity issue is potting failure under the thermal stress for membrane modules experiencing sudden temperature swing, such as pumping hot cleaning chemical solutions quickly into the modules under cold ambient temperature. The thermal stress is caused by uneven heating of different parts of potting and can cause cracks in potting layer.

Another cause for integrity breach is incidental exposure of membranes to incompatible chemicals and/or environment. The plant operator needs to make sure that membranes operate in the proper pH and temperature ranges, and chemical cleaning of membranes needs to follow the instructions of membrane suppliers.

Long-term Wear and Tear

Low-pressure membranes need to endure repeated mechanical stress in their service life from operations such as filter backwash and air scrubbing. In addition, membranes may also be exposed to chemicals (either from water treatment or from cleaning) over their service life. Both mechanical stress and chemical exposure are of chronic nature, and their long-term effect on membranes may be cumulative.

Mechanical Stresses Upon Hollow-Fiber Membranes

The mechanical stresses applied to membranes are of cyclic nature as membrane filters have repeated filtration–backwashing cycles. It has been shown that cyclic load tends to be more deteriorative than monotonic load as the former is more likely to cause

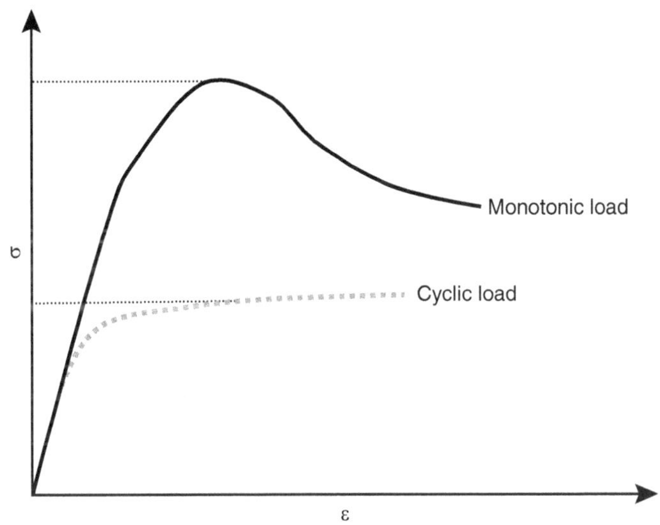

Figure 2-7. Impacts of different loads on stress–strain relations

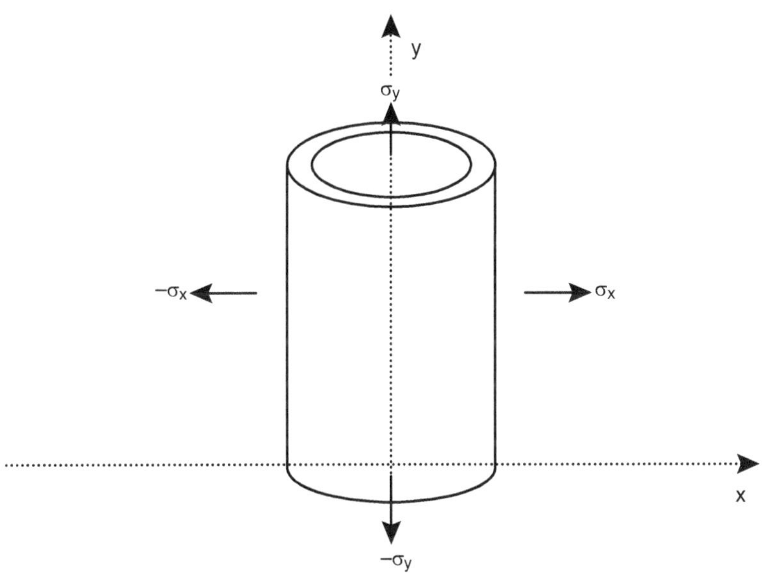

Figure 2-8. Major stresses applied to hollow-fiber membranes (σ_x = radial stress; σ_y = longitudinal stress, i.e., tension)

material fatigue fracture or "aging" (DeVries and Nuismer 1985). The stress–strain curve for polymeric materials shows a much lower yield stress when the polymers are subjected to cyclic loading, as illustrated conceptually in Figure 2-7.

To understand this issue better, the major mechanical stresses acting on hollow-fiber membranes should be analyzed, as depicted in Figure 2-8. The major stresses can be expressed in two components—a radial component, σ_x (i.e., the direction

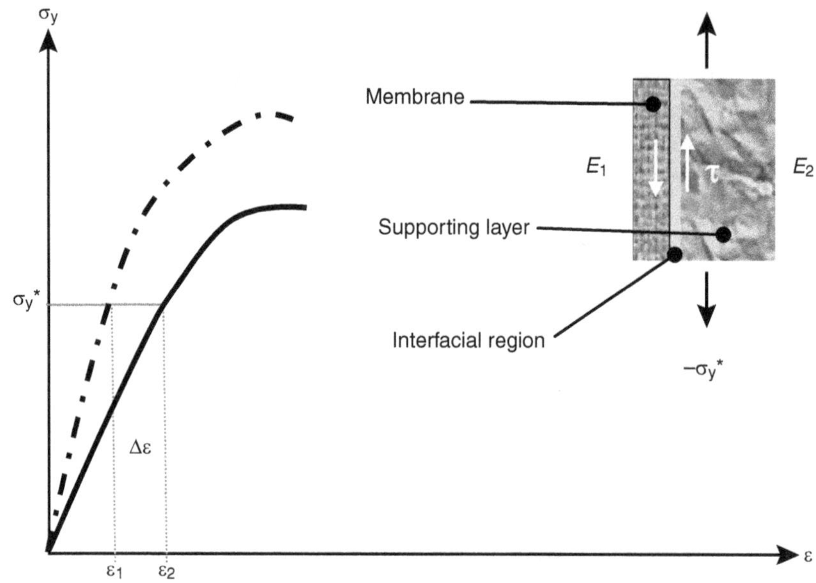

Figure 2-9. Shearing stress in interfacial region of multi-cloth hollow-fiber membrane caused by the given tension σ_y* (σ_y = tension; ε = elongation; E_1 = Young's modulus of membrane; E_2 = Young's modulus of supporting layer; τ = interfacial shearing)

perpendicular to the membrane surface), and a longitudinal component, σ_y (i.e., the direction parallel to membrane surface). The stress in the longitudinal direction occurs when water flows along the hollow fibers as in both normal filtration and backwash. In addition, tension during the air-assisted backwash of hollow fibers can occur as ends of fibers are fixed by potting. A radial stress can occur when water flows through membranes during the normal filtration cycle or backwash. As a result of radial stress, the hollow fibers experience repeated expansion and compression in filtration/backwash cycle. This radial stress also generates a shear toward hollow fibers on the face of potting.

For composite-membrane hollow fibers, the stress in radial direction exerts a pressure that "pushes" functional membrane layer and supporting layer apart. The stress in longitudinal direction exerts a tension on the hollow fiber. Because of the difference in elasticity of membrane and supporting materials, this tension would create a shearing stress on the interfacial region as illustrated by Figure 2-9.

In Figure 2-9, when the hollow fiber is subjected to a longitudinal tension, σ_y^*, the strains for the membrane medium and the supporting layer are given by

$$\varepsilon_1 = \sigma_y^*/E_1 \quad \text{(Eq 2-1)}$$

and

$$\varepsilon_2 = \sigma_y^*/E_2 \quad \text{(Eq 2-2)}$$

Where

E_1 and E_2 = modulus of elasticity of the membrane and the supporting layer, respectively

ε_1 and ε_2 = strains for the membrane and the supporting layer, respectively

$\sigma_y{}^*$ = longitudinal tension

As a result of the difference in elasticity of the membrane medium and the supporting layer, the strains caused by the tension would be different for the two materials.

$$\Delta\varepsilon = \varepsilon_2 - \varepsilon_1 = \sigma_y{}^* \left(\frac{1}{E_2} - \frac{1}{E_1} \right) \quad \text{(Eq 2-3)}$$

Where

$\Delta\varepsilon$ = differential expansion between the membrane and the supporting layer

The differential expansion of neighboring materials caused by tension then creates a shear stress on the interfacial region:

$$\tau = G\Delta\varepsilon = G\sigma_y{}^* \left(\frac{1}{E_2} - \frac{1}{E_1} \right) = \frac{G\sigma_y{}^*(E_1 - E_2)}{E_1 E_2} \quad \text{(Eq 2-4)}$$

Where

τ = shear stress between the membrane and the supporting layer

G = shear modulus at the membrane–supporting layer interface

Equation 2-4 indicates that the shearing force is proportional to the aptitude of the longitudinal tension, the difference in modulus of elasticity, and shear modulus, G, of the interfacial region, and inversely proportional to the products of modulus of elasticity of membrane and supporting materials. As previously discussed, both longitudinal and radial stresses can contribute to the delaminating of composite membranes.

Chemical Degradation of Polymeric Membranes

In addition to mechanical stress, membranes are subjected to periodic chemical cleaning as they are fouled. They are also often in contact with certain water treatment chemicals (e.g., acids, bases, and oxidants). Therefore, chemical degradation of polymeric membrane materials can also occur. If the stress levels exceed the endurance of membrane materials, fractures of membranes and integrity breach may occur.

For example, membranes made from cellulose acetates (CAs) tend to have relatively narrow pH range. Beyond this range, hydrolysis of membrane media would accelerate considerably and membranes would deteriorate prematurely. Similarly, the chemical compatibility of PVDF membranes to alkaline is an issue. The PVDF membranes with low crystallinity have low tolerance to strong alkaline such as caustic soda. This is because PVDF can undergo dehydrofluorination reaction that makes the membranes more brittle and breakable under mechanical stresses.

Chemical degradation refers to any type of modification of a polymer chain involving the main-chain backbone or side groups, or both, that adversely affects the physical properties of polymeric membranes caused by contact with chemical reagents. In general, polymeric membranes are subject to two types of chemical degradation: *hydrolysis* and *oxidation*.

Hydrolysis is a chemical process in which decomposition of a compound occurs as the result of its reaction with water. The hydrogen and hydroxide ions split from a water molecule are added to each of the products of hydrolysis reaction. Hydrolysis can also be considered as the reverse reaction of polymerization by condensation. Hydrolysis may occur under neutral, acid, or alkaline conditions.

Generally, neutral and acid hydrolyses are similar, whereas alkaline hydrolysis is quite different. Hydrolytic degradation of polymers may involve main-chain scission because of functional inter-units in condensation polymers (e.g., amide and ester linkages in polyamides and polyesters, respectively) and/or side groups. The importance of hydrolysis stems from the breakdowns of physical properties (e.g., tensile strength) in polymers. The hydrolysis of CA membranes occurs when 1,4-b-glycoside binds connecting thousands of saccharide (i.e., sugar) units are broken down. This is why CA membranes typically have a narrower pH range than membranes made with other media.

Oxidative degradation of polymers is more complicated because it is affected greatly by the changes in the reaction conditions (e.g., the presence or absence of inhibitors, the presence or absence of metal catalysts, etc.). Generally, it involves an initiation step in which certain structures within polymeric materials acting as a free-radical initiator undergo unimolecular or bimolecular decomposition and generate free radicals. The initiation step is followed by propagation step in which free radicals react with polymeric materials to generate organic peroxides (RO_2) and/or hydroperoxides (RO_2H).

Finally, the reaction process undergoes termination steps in which products of the reaction are generated via decomposition and/or recombination of intermediate products. As such, oxidative degradation of polymers tends to be structurally selective, i.e., certain structural features on the polymers are more vulnerable to the chemical attacks from oxidants than the polymer as a whole.

The symptoms of chemical degradation may include one or more of the following changes in a material: hardening, embrittlement, softening, cracking, crazing, discoloration or alteration of certain properties such as rejection efficiency, and weakening of mechanical strength. Chemical degradation, compounding mechanical stresses in the operation, can be very detrimental to service life of membranes and cause integrity breach as membranes rupture under the stress.

SIGNIFICANCES OF CONDUCTING MEMBRANE INTEGRITY TESTING

Regulatory Compliance

The objective of applying low-pressure membrane technologies in drinking water is to effectively remove pathogenic microbial contaminants and produce microbiologically safe drinking water. Membrane integrity testing is a tool used for this objective. As discussed previously, the mechanism for low-pressure membranes is size exclusion, as long as integral low-pressure membranes are able to completely retain the pathogens above their size of exclusion. However, the retaining characteristics have to be confirmed by integrity testing. In this sense, membrane integrity testing can provide a positive verification on the process efficiency on routine basis—a unique feature among water treatment processes.

The recent regulation Long-Term 2 Enhanced Surface Water Treatment Rule (LT2ESWTR) in the United States specifically specifies that membrane processes conduct integrity testing as part of compliance requirement. Therefore, for low-pressure membrane systems intended for regulatory compliance, conducting integrity testing is a necessity.

According to the LT2ESWTR, integrity testing has to be conducted at two levels: manufacturing and installed system. At the manufacturing level, integrity testing is used as a quality assurance and quality control (QA/QC) tool. Each filter module has to pass a *nondestructive performance test* (NDPT) in which a *quality control release value* (QCRV) is established in accordance with the performance criteria specified in the rule. This factory-instituted test would ensure the performance and integrity of the individual filter module shipped. After a membrane system is installed and put into operation, integrity testing has to be conducted routinely in accordance with the performance criteria specified in the rule. The removal credit can only be granted based on the result of the testing. A detailed account on the regulatory requirements and performance criteria specified in LT2ESWTR is presented in chapter 3. For the low-pressure membrane systems installed in compliance with the LT2ESWTR, meeting the requirements for membrane integrity testing is the single most important issue.

Quality Control for Manufacturers

As stated earlier, LT2ESWTR requires that a QCRV be established at the factory. The manufacturing process consists of the making of hollow fibers and assembling of filter modules. The process of making hollow fiber is a semi-batch process, as each batch of solution of base materials is prepared in a coagulation bath. Consistency of quality of hollow-fiber membranes depends on the batch-to-batch consistency of base materials and processing conditions for fiber spinning. Establishing quality controls of hollow fibers (e.g., rejection, permeability, and mechanical strength) before assembling them into filter modules is important not only to ensure product quality but also to prevent waste of labor and materials in assembling filter modules. After assembling fibers into the module, each individual module shall be tested again to meet QCRV for integrity and permeability criterion. Frequent occurrence of substandard products points to systematic deficiency in the production that needs to be overhauled. In this sense, integrity testing is a part of a broad production QA/QC program acting both for quality control and productivity enhancement. The integrity testing of filter module is another important step in production QA/QC as it is the final step to ensure that products meet quality specification.

Monitoring and Diagnosis Tool for Plant Operators

A well-designed and well-operated low-pressure membrane plant typically should not have excessive integrity breaches. When an excessive breach in integrity occurs, it may point to problems in either design or operation. The word *excessive* is a somewhat inaccurate and objective modifier because its definition varies among membrane suppliers. However, there should be references from other similar plants, or even specifications provided by the membrane supplier, to define what constitutes excessive. The occurrence of integrity breaches is a function of the service time as a result of long-term wear and tear.

Because many low-pressure membrane plants were commissioned in the past few years and because there are not sufficient experiences in long-term performance with respect to the history of integrity breaches, it may be difficult to establish the benchmark for what is excessive at "near the end" of the service life. Nevertheless, a significant increase in occurrence of integrity breaches may signal that something is wrong. If this increase ever occurs, it should warrant an in-depth investigation into the causes of the problem.

3

IMPORTANT CRITERIA OF MEMBRANE INTEGRITY TESTING

INTRODUCTION

In principle, membrane integrity can be defined as *wholeness*—the state of being defect free. However, this definition hinges on what an integrity testing method can actually detect. Therefore, before it can be determined whether a membrane system is integral, the criteria for membrane integrity testing should be established. That is, the following questions need to be answered:

1. What is to be detected (resolution)?
2. How can the impact of an integrity breach be determined (sensitivity)?
3. How often should the integrity of a given membrane system (testing frequency) be verified?
4. How can it be assured that the detection is real, not a false alarm (reliability)?
5. If evidence of an integrity breach is detected, how soon can the origin be traced (traceability)?
6. How is an integrity-testing method or procedure to be implemented (implementability)?

Those criteria determine how a membrane integrity testing method is to be evaluated (Liu et al. 1999; Liu and Schaefer 2002). Those concepts were adapted in the request for proposal (RFP) for a project funded by the Awwa Research Foundation (now known as the Water Research Foundation) in 2000 with minor modifications. (The term *identifiability* was changed to *traceability.*) In its *Low-Pressure Membrane Filtration for Pathogen Removal: Application, Implementation, and Regulatory Issues*, the US Environmental Protection Agency (USEPA) listed three criteria ("a theoretical, a practical, and an operational criterion") that must be satisfied to ensure that microbial risk is minimized (USEPA 2001):

1. Classification of pore size (theoretical criterion)
2. Demonstration of particle removal (practical criterion)
3. Verification of membrane integrity (operational criterion)

The first criterion is the precursor for the resolution criterion in the Long-Term 2 Enhanced Surface Water Treatment Rule (LT2ESWTR) for direct integrity testing; while the second and the third criteria form the basis for the requirements of challenge tests and integrity verification for the membrane systems used for compliance with the rule. In addition, the 2001 USEPA document summarized several outstanding issues raised by states. Among them, "reliability, sensitivity and transparency of integrity testing methods" and "increased understanding of the relationship between integrity testing results and a breach in integrity and risk of microbial passage" were mentioned. These issues became the basis for the regulatory framework for integrity testing of membrane systems in the LT2ESWTR, in which resolution, sensitivity, and testing frequency are deemed the three performance criteria for the direct integrity testing of membrane systems for rule compliance.

The important criteria for integrity testing are discussed in the following sections.

RESOLUTION

Resolution is defined as the smallest membrane integrity breach that can be detected. The finer the resolution, the smaller the size of an integrity breach that is detectable by a given testing method. If a membrane is integral (defect free), anything larger than the largest pore of the membrane would be absolutely retained. (For simplicity's sake, it is assumed that no deformation of particles is occurring). When a membrane has a breach that results in a larger opening than its largest pore, its separation characteristics then change, as the breach becomes the largest "pore." The membrane is no longer the absolute barrier for a contaminant whose dimension is greater than the largest pore of the integral membrane but smaller than the opening.

There is a probability that a contaminant smaller than the dimension of the breach could enter the filtered stream through the breach. This has obvious health implications when the membrane is used to filter drinking water. Therefore, the real question is whether the resolution of an integrity testing method is sufficient to allow detection of the potential passage of the contaminants targeted for removal by the membranes. If the answer is no, then there is no assurance that the membrane can retain the contaminants as it should, regardless of the rating of the membrane's pore size and regardless of whether the membrane is actually integral. Following this logic, the retaining ability of a given membrane is as good as the resolution of which membrane integrity testing can provide. From the regulatory perspective, in which minimizing public health risk is the focus, this line of thinking is justified, even though the risk is merely theoretical. Conversely, if the contaminant targeted for removal by membranes is defined, one can ask under what testing conditions the required resolution criterion can be met. Therefore, the concept of resolution is not only theoretical but also practical.

One of the factors contributing to the exponential growth of low-pressure membranes in drinking water treatment might have been the discovery in later 1980s and early 1990s of disinfection-resistant pathogens (e.g., *Giardia* and *Cryptosporidium*) and the subsequent regulations to reduce the risks of microbial contamination. Therefore, it makes sense to set resolution criterion of 3 μm (the lower end of the size range for *Cryptosporidium*) as *Cryptosporidium* is the targeted pathogen in the LT2ESWTR.

For ultrafiltration (UF) membranes, the resolution criterion raises an interesting question. Even though UF membranes have demonstrated the ability to remove viruses, current integrity testing methods for routine integrity verification are unable to get the resolution down to the size of viruses (presumably 0.025 μm). Nor does a practical method that can be used for the purpose of quality assurance and quality control (QA/QC) during the manufacturing process exist to verify that membranes are defect free. As a result, a quality control release value (QCRV) cannot be established for nondestructive performance test (NDPT) as required in LT2ESWTR (USEPA 2006a). Because of a lack of practical methods to verify the resolution of membranes, the issue of whether credit for viruses should be granted is being debated within the industry (Alspach 2007).

Different approaches regarding virus credit are used by state regulatory agencies. For example, California's Department of Public Health takes the approach of granting virus removal credit based on the review of results of microbial challenge tests. This approach seems quite different from the regulatory framework established in LT2ESWTR, whose philosophy seems to be "no verification, no credit." Some states limit the virus removal credit by as much as several logs below the log removal value (LRV) demonstrated by the challenge tests for safety. Because viruses can be easily inactivated by disinfection and all utilities have some forms of disinfection, the low virus removal credit granted to low-pressure membranes by the state regulatory agencies typically does not cause problems for regulatory compliance.

Detailed discussions of the integrity testing methods based on fluid flow will be presented in chapter 5.

SENSITIVITY

Sensitivity is defined as the maximum LRV that can be reliably verified by an integrity test. If resolution is to address "how small," then sensitivity is to address "how much." According to LT2ESWTR, the removal credit for *Cryptosporidium* is granted based on the lower value from (a) product-specific microbial challenge test and (b) site-specific verification through direct integrity testing. This is consistent with the philosophy of "no verification, no credit." Because pore sizes of low-pressure membrane filters used in drinking water applications are at least one order of magnitude smaller than the size of *Cryptosporidium*, microbial challenge tests performed on integral membranes can yield removal down to the method detection limit (MDL) of those pathogens regardless of the challenge level (the concentration of pathogens in the feed). Practically, therefore, the removal credit is to be governed by the LRV verified by membrane integrity testing.

In other words, sensitivity of integrity testing is actually determining the removal credit that can be granted to a given membrane system. As such, the sensitivity of integrity testing for a given membrane system as measured by LRV needs to be no less than that required for compliance with the designated regulations.

Mathematically, the log removal value is expressed as the following:

$$\mathrm{LVR} = \log \frac{C_0}{C} \qquad \text{(Eq 3-1)}$$

where C_0 and C = concentration of the targeted contaminant (e.g., *Cryptosporidium* oocysts) in the feed and filtrate streams, respectively.

For an integral system for which the target contaminant in the filtrate approaches method detection limit, the equation to calculate LRV becomes:

$$\mathrm{LVR} > \log \frac{C_0}{\mathrm{MDL}} \qquad \text{(Eq 3-2)}$$

where MDL = method detection limit for the target contaminant.

Note that ">" instead of "=" is used in Equation 3-2. This is because MDL is typically expressed as "not greater than" (i.e., ≤) a certain value. Therefore, when Equation 3-2 is used for LRV calculation, the result is expressed as a minimum value, i.e., the LRV that an integral membrane system at the least can achieve.

Note also that a common mistake in calculating LRV for an integral system is simply to use a logarithm for the feed concentration instead of using Equation 3-2. Using the logarithm for the feed concentration is only accurate if the MDL for the contaminant is 1. If the value of MDL is less than 1, an LRV calculation using the logarithm for the feed concentration could lead to an erroneous result—a lower LRV for an integral membrane system than that of a membrane system with a small membrane breach. For example, assume that the MDL for a given contaminant is 0.01 and the measured concentrations of the contaminant in feed and filtrate of an integrity-compromised system are 10,000 and 0.5, respectively. LRV calculated using the logarithm of the feed concentration for the integral membranes is 4-logs, while LRV for the integrity-compromised system is 4.3-logs. The correct LRV for an integral system should be > log (10,000/0.01) = 6.0-logs. Again, the LRV value here should be viewed as the removal efficiency that the integral membrane system can at the least achieve, as the value is dependent of the feed concentration.

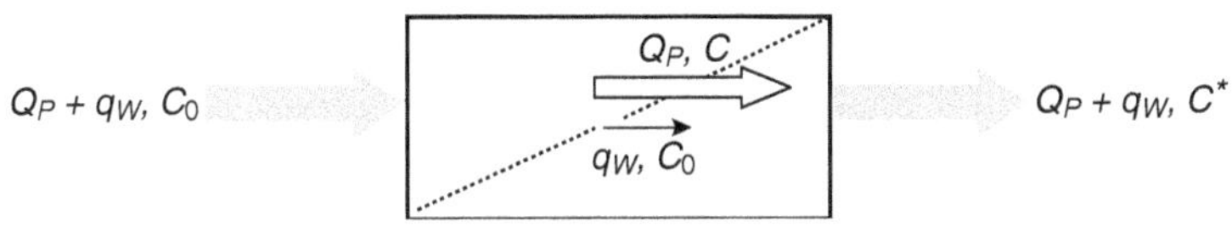

(Q_P = filtrate flow; q_W = bypass flow; C_0 = concentration in feed; C = concentration in filtrate of intact membranes; C^* = concentration in filter effluent)

Figure 3-1. Mass balance of membrane filtration system with an integrity breach

For a membrane system with a breach (e.g., a broken hollow fiber) depicted in Figure 3-1, mass balance for a targeted contaminant can be expressed by the following:

$$(Q_P + q_W)C^* = Q_P C + q_W C_0 \qquad \text{(Eq 3-3)}$$

Where

Q_P and q_W = filtrate from intact membranes and bypass flow from the membrane breach, respectively

C^*, C, and C_0 = concentrations of the targeted contaminant in filter effluent, filtrate of intact membranes, and bypass flow, respectively

Rearrange Equation 3-3 and we have:

$$C^* = \frac{Q_P}{Q_P + q_W}C + \frac{q_W}{Q_P + q_W}C_0 \qquad \text{(Eq 3-4)}$$

Assume the concentration of the targeted contaminant in the filtrate of intact membranes C approaches zero, and filtrate from intact membranes Q_P is much greater than bypass flow q_W (typical for the membrane breach of a single broken fiber), Equation 3-3 can be approximated by the following:

$$C^* \approx \frac{q_W}{Q_P}C_0 \qquad \text{(Eq 3-5)}$$

Rearrange (3-5):

$$\frac{C_0}{C} \approx \frac{Q_P}{q_W} \qquad \text{(Eq 3-6)}$$

Equation 3-6 states that the concentration of the targeted contaminant is essentially the result of dilution of the contaminants in bypass flow by filtrate from intact membranes.

In certain hydraulic configurations of membrane systems, contaminants are concentrated upstream of membranes. Two examples of contaminant concentration are illustrated in Figure 3-2. This concentrating effect can be expressed by volume concentration factor (VCF):

$$\text{VCF} = \frac{C_M}{C_0} \qquad \text{(Eq 3-7)}$$

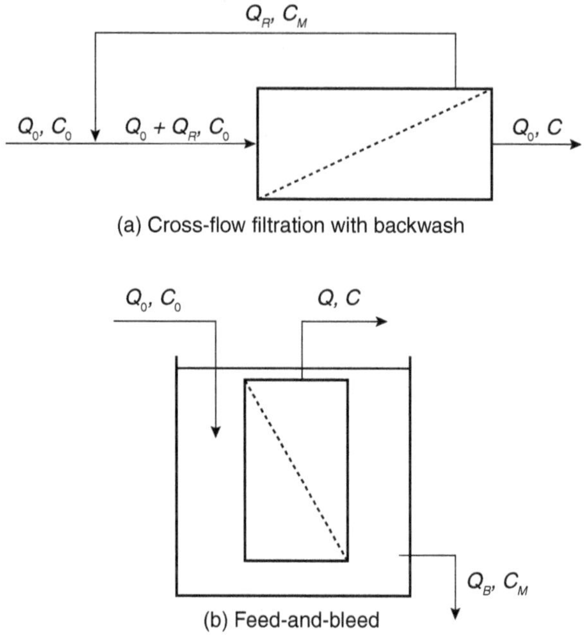

(Q_0 = feed flow; Q_R = recycled flow; Q = filtrate flow; C_0 = concentration in feed; C_M = concentration in upstream of membrane; C = concentration in filter effluent; Q_B = bleed flow)

Figure 3-2. Examples of contaminant concentration in low-pressure membrane systems

Because of the concentration effect, contaminant concentration upstream of the membrane (C_M) is greater than or at least equal to that in the feed stream (C_0). Therefore, VCF must be ≥1.0. For a membrane system operating in a dead-end mode, 100 percent of the feed passes through membranes during filtration phase. Therefore, the concentration at upstream membranes is equal to that of the feed stream, i.e., VCF = 1. Alternatively, a membrane system operating in a cross-flow mode concentrates contaminants in the system because only a portion of the feed stream is passing through the membranes and the rest is circulated back to the upstream side of membranes, i.e., VCF > 1.

Assuming that the target contaminant is suspended in the water, VCF for various operating modes and hydraulic configurations can be derived via mass balance. A detailed work on the derivations of VCF for different operating modes of various membrane systems can be found in chapter 2 of the *Membrane Filtration Guidance Manual* (USEPA 2005).

Table 3-1 summarizes the equations for calculating VCF for various hydraulic configurations. The typical values of VCF derived from equations in Table 3-1 are given in Table 3-2. As indicated by the values in Table 3-2, VCF has a significant impact on LRV. For example, for membrane systems operated at cross-flow mode, LRV would be lowered by 0.6- to 1.3-logs for VCF values ranging from 4 to 20.

Table 3-1. Summary of equations for VCF calculation for various systems (adapted from Table 2.4 of USEPA *Membrane Filtration Guidance Manual*)

Membrane System	Typical Process	Equations for VCF Calculation	
		Average Value	Maximum Value
Dead-end	MF/UF	1	1
Small-volume cross-flow ($Q_B t_B < V$)	MF/UF without feed tank	$0.5\left(\frac{t}{\tau}\right)$	$\left(\frac{t}{\tau}\right)$
Large-volume cross-flow ($Q_B t_B > V$)	MF/UF with feed tank	$\frac{Q_0 t}{Q_B t_B}\left[1 - \frac{t}{2\tau \frac{Q_0 t}{Q_B t_B}}\right]$	$\frac{Q_0 t}{Q_B t_B}$
CSTR without backwash	MF/UF with bleed	$\frac{1}{1-R}$	$\frac{1}{1-R}$
CSTR with backwash	MF/UF with bleed	$\frac{1}{1-R}\left[1 - \left(\frac{\tau}{t}\right)\left(1 - \exp\left(\frac{-t}{\tau}\right)\right)\right]$	$\frac{1}{1-R}\left[\left(1 - \exp\left(\frac{-t}{\tau}\right)\right)\right]$

Notes:

VCF = volume concentration factor
Q_B = backwash flow rate
t_B = backwash duration
V = hold-up volume of the system
Q_0 = feed flow
t = filtration duration
τ = system turnover time
R = system recovery (%)
MF = microfiltration
UF = ultrafiltration
CSTR = continuously stirred tank reactor

Table 3-2. Typical range of VCF values for various operating modes and hydraulic configurations (adapted from Table 2.3 of USEPA *Membrane Filtration Guidance Manual*)

Membrane System/Hydraulic Configuration	Range of VCF Values
Dead-end	1
Cross-flow (including small and large cross-flow)	4–20
CSTR (with or without backwash)	4–20

VCF = volumetric concentration factor
CSTR = continuously stirred tank reactor

Because VCF values in Table 3-1 represent ideal systems and an actual membrane system may deviate from those ideal systems, the *Membrane Filtration Guidance Manual* allows system-specific demonstration to establish VCF value. For a specific system, VCF value can also be determined experimentally by applying Equation 3-7. The following is an example of determining VCF value calculated using data from a microbial challenge test on a production membrane system in which about 10 percent the feed stream is circulated in a cross-flow fashion (Sethi et al. 2004). The samples of feed, concentrate, and filtrate streams (25 samples each) were taken for analysis of challenge reagent, *Bacillus subtilis* spores (≈1 μm). VCF values were calculated via Equation 3-7 by using paired data of spore counts in the feed (C_0) and the concentrate (C_M) streams. A percent distribution of VCF values was constructed as shown in Figure 3-3. As indicated in the distribution curve, the 75th percentile value is very close to 1 (1.08), while the 99th percentile value is 1.60. The VCF distribution of this particular system in the example indicates that its operating mode is very close to that of a dead-end system. VCF value determined empirically is system-specific and only applicable to the systems of the same hydraulic configuration and operating conditions.

For membrane systems with VCF > 1, C_0 in Equation 3-1 needs to be replaced by C_M, that is:

$$\text{LRV} = \log \frac{C_0}{\text{VCF} \times C} \qquad \text{(Eq 3-8)}$$

Substitute Equation 3-6 into Equation 3-8:

$$\text{LRV} = \log \frac{Q_P}{q_W \times \text{VCF}} \qquad \text{(Eq 3-9)}$$

As indicated in Equation 3-9, LRV is the result of dilution. Therefore, sensitivity is really a matter of detecting a small bypass flow relative to a large filtrate flow. Sensitivity can be determined through either Equation 3-1 or Equation 3-9, depending on the methods of integrity testing. For the systems where target contaminants (or their surrogates) are actually measured in feed and filtrate streams, LRV calculation using Equation 3-1 or Equation 3-2 (if contaminant concentration in the filtrate stream is below MDL) is straightforward. For the systems where integrity testing methods other than measuring contaminant concentration are used, the matter to determine sensitivity becomes determining the bypass flow for given operating conditions. The approach is to correlate the testing parameter(s) in an integrity test to the bypass flow of the given system. Detailed discussion on the basis of such an approach and development of equations are given in chapter 5.

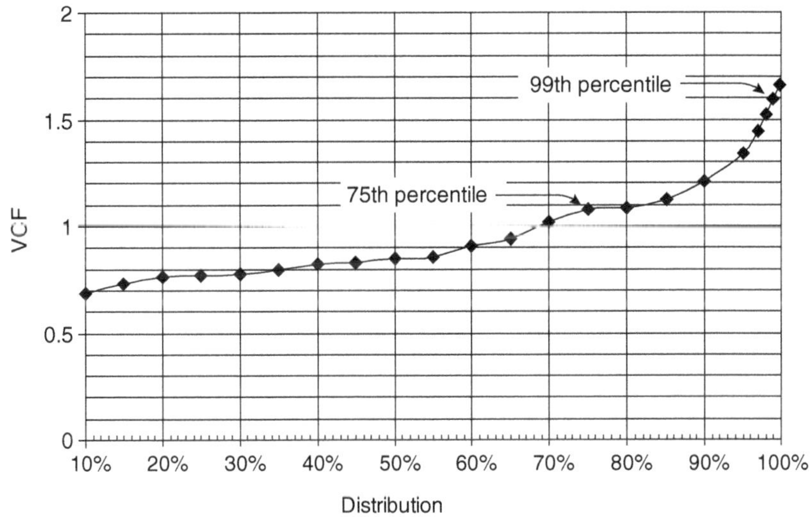

Figure 3-3. Statistical distribution of VCF value based on microbial challenge test

TESTING FREQUENCY

Ideally, integrity monitoring is continuous, so that the operator knows a membrane integrity breach at the moment it occurs. Unfortunately, the continuous integrity monitoring methods currently available are neither sufficiently sensitive nor reliable for such a task. The most sensitive and reliable integrity monitoring methods are typically not continuous, and almost all of them require the membrane system to be off-line when the test is performed.

This raises the question of how often integrity testing should be performed. LT2ESWTR specifies that testing frequency for membrane systems should not be less than once per day—a compromise between risk and practicality. Although the rule permits regulatory agencies with primacy to determine the testing frequency as they see fit, it does not provides guidance in determining the testing frequency. As a result, there are discrepancies among the states regarding the required testing frequency. According to a survey conducted by AwwaRF, nearly 60 percent of membrane plants are required to perform (direct) integrity testing once per day, and 25 percent are required to perform testing more frequently (Adham et al. 2005).

At the heart of the issue, it is a balance of risk of compromised safety and quality of the finished water from loss of membrane integrity versus the practical consideration of the effect on plant productivity of the downtime required to perform integrity testing. Therefore, it is possible to apply a risk-based approach to assessing testing frequency (Liu and Schaefer 2002; Liu 2006).

DEFINE THE RISK

From the perspective of public health, a risk is the product of the probability of the exposure of the public to elements harmful to human health and the consequence of such exposure. In a quantitative expression (Hellenbeck 1986):

Risk (R) = Probability of Exposure (P) × Consequence of Exposure (C) (Eq 3-10)

Equation 3-10 indicates that both the probability of exposure and the consequence of the exposure contribute to risk. That is, a low-probability event with a dire consequence (e.g., a category 5 hurricane landing on a major metropolitan area in the United States) may be as risky as a more frequently occurring event whose destruction is less dramatic (e.g., an outbreak of mild influenza that causes many people to become sick). If the same concept is used to assess the risks of membrane integrity breach, Equation 3-10 becomes:

Risk = Probability of Occurrence (of Breach) × Consequence (of Breach) (Eq 3-11)

As indicated in Equation 3-11, the risk can be expressed as the occurrence of an integrity breach multiplied by the consequence of such a breach. As long the two terms on the right side of the equation can be quantified, the risk of membrane integrity breach can be assessed.

Evaluating the frequency of integrity testing has three parts:

1. *Occurrence analysis:* frequency of occurrence, mode of breach, and possible causes of breach
2. *Consequence evaluation:* the impact of integrity breach on water quality
3. *Risk assessment based on occurrence and consequences:* decision-making on testing frequency

Determine the Probability of Occurrence of Membrane Integrity Breach

Factors Contributing to Membrane Integrity Breach

Many factors contribute to the occurrence of a membrane integrity breach. Those factors can be classified into three categories:

- Materials and membrane/module configuration
- Manufacturing process and QA/QC procedures
- Operating environments

Most MF/UF membranes used in drinking water treatment are polymeric membranes. The most commonly used membrane materials include polyvinylidene fluoride (PVDF), polyethersulfone (PES), polysulfone (PS), polypropylene (PP), and cellulose acetate (CA). According to an American Water Works Association Research Foundation (Adham et al. 2005) survey, more than half of the installed capacity in million gallons per day (mgd) of different MF/UF membrane materials is PVDF. Other membrane materials (PES, PS, PP, and CA) range from 8 percent to 15 percent of the installed capacity. The dominance of PVDF membranes may relate to the general characteristics of high chemical resistance and mechanical strength.

In addition to membrane materials, membrane configuration also plays an important role. Most of MF/UF membranes are mono-cloth membranes. That is, they are made of one homogeneous material. However, some membranes are multi-cloth—two or more different polymers acting as active membrane and supporting substrate, respectively. In the latter case, different materials are "glued" together through interfacial polymerization. This type of configuration is used mostly in nanofiltration (NF) and reverse osmosis (RO) membranes (e.g., thin-film composite membranes) where membranes do not experience mechanical stress, unlike MF/UF membranes that have to withstand repeated backwash or air-assisted backwash. When the multi-cloth membranes are under mechanical stress (e.g., radial and longitude tension), the interfacial polymerized layer would be under either a normal or a shearing stress (see chapter 2 for detailed analysis). The active membrane surface and the underlying supporting substrate separate when the stress exceeds strength of the interfacial layer. As a result, the membranes integrity is lost even if hollow fibers are not physically broken.

Manufacturing process can also enhance or weaken the chemical strength and mechanical strength of membranes. It is known that polymers with highly crystalline structure have high mechanical strength and chemical resistance. Manufacturing processes such as heating, cooling, and annealing all affect the degree of crystallization of polymers (Kumar and Gupta 1998). This is why the same membrane materials can exhibit different mechanical strength and chemical resistance. Another important issue in reducing occurrence of membrane integrity breach is production QA/QC procedures, which prevent defective membranes from being installed. Under LT2ESWTR, all membrane manufactures are required to conduct a nondestructive performance test (NDPT) for all the membrane modules that are not be subjected to a challenge test. Membrane manufacturers are also required to establish a quality control release value (QCRV) to ensure the minimum performance standard of products shipped for installation (USEPA 2006a).

Operating environments after the membranes are installed can also be a very important factor affecting the occurrence of membrane integrity breach. Extreme temperatures, out-of-spec operating conditions, improper chemical cleaning procedures, and the number of years in service can all affect the occurrence of membrane integrity breach.

Finally, the sensitivity and reliability of membrane integrity tests also need to be considered. Even if tests do not affect occurrence of breach, they affect the detection of occurrence. In reality, only when the integrity breaches are detected are they counted as occurrences. Therefore, the sensitivity and reliability of integrity testing method have to be taken into consideration when determining occurrence.

Determining the Probability of Occurrence

Because many factors can contribute to occurrence, determining occurrence can be difficult. Two practical approaches can be applied to determine the occurrence of a membrane integrity breach: (1) durability tests and (2) historical data about already installed membrane systems.

A durability test, also called a *fatigue test*, is a standard test for a system or a component that is expected to have a long working life. Ideally, a durability test mimics the conditions under which systems are conducted. For example, the microfiltration and ultrafiltration systems used in drinking water treatment undergo frequent backwash and routinely are exposed to chemicals for cleaning. Therefore, those systems experience both mechanical and chemical stresses, and the durability tests must be designed to consider both factors. Because filter modules are expected to have a relatively long life (5 to 10 years), durability tests are typically accelerated stress tests to compress the time frame needed to simulate the filter modules' life cycle.

Tests are typically performed using specially designed automatic testing rigs. The actual conditions of filter backwash are replicated at much more frequent intervals (e.g., at a ratio of 10:1 or more). For chemical stress, the accelerated fatigue tests can be performed by increasing concentrations of the chemicals and the exposure time to simulate chemical cleaning. Using a linear exposure model, the equivalent exposure (the product of exposure time and concentration) can be extrapolated to the actual exposure conditions. During both mechanical and chemical fatigue tests, samples must be taken periodically to test for fatigues and material deterioration.

One advantage of conducting fatigues tests is to obtain a histogram of material deterioration over the lifetime of filters. Based on the histogram, the probability of membrane breach can be constructed as a function of service time. This is significant because it generates the information on which to base expectation of a rising trend of membrane breaches. Thus, closer monitoring (i.e., membrane integrity testing) can be planned ahead of time. Because fatigue tests are conducted under standard and controlled conditions, the results are reliable and universally applicable. The disadvantage of this approach is that running fatigue tests is time-consuming and expensive.

The second approach to establishing the probability of occurrence is to estimate using historical data of the membrane systems currently in use. The reports on incidents of membrane integrity breach are rare and results are mixed (Naismith 2005; Adham et al. 2005; Panglisch et al. 1998). In one instance, it was reported that three broken fibers were found at the startup, and six more during five years of operation at a 7.8-mgd plant (Naismith 2005). The most comprehensive data are from a report published by American Water Works Association Research Foundation (AwwaRF) describing a study in which 36 MF/UF plants were surveyed and the incidence of fiber breakage was tallied (Adham et al. 2005). Normalized on the plant capacity, the median number of fiber breakage incidents per year per million gallons per day was 1, with a 90th percentile of 5. The results also indicated that during the first three months from startup, the fiber breakage incidence was slightly higher than the annual average. In terms of the number of compromised fibers per incident, the median value was 3, with a 90th percentile of 10.

About 10 percent of plants surveyed appeared to have a significantly higher number of both fiber breakage incidents and number of compromised fibers per incident. Because of system- and site-specific conditions, historical data only serve as a reference and do not necessarily represent all plants. However, by reviewing historical data, one may evaluate the relative performances of various systems regarding membrane quality and durability from perspective of integrity.

When reviewing the historical data, it is important to conduct failure analyses, including the frequency of occurrence, the mode of failures, and the causes of failures. The purpose of failure analysis is to identify whether the failure is caused by manufacturing issues or from improper construction, installation, and/or startup. The reason to make such a distinction is to more accurately assess long-term occurrence of membrane integrity breaches by isolating the incidents caused by temporary events.

Determine Consequence of Integrity Breach

The historical data concerning membrane integrity breaches indicate two modes of breaches: pinhole and partially cut fibers. The essential questions are

- How do those integrity breaches affect the water quality of a membrane system?
- What is the risk to the public when the breaches indeed occurred?

This section will present an approach to assess the impact of a single broken fiber on water quality. This approach includes the following steps:

1. Establish bypass flow from a single completely broken fiber. This flow is greater than the bypass flow through a pinhole or a partially cut fiber and therefore is the most conservative estimate of bypass flow from a single damaged fiber.
2. Establish the log reduction value (LRV) of membrane systems.
3. Assess the risk of the membrane systems with defects.

Bypass Flow From a Single Broken Fiber

Bypass flow from a single broken fiber depends on the configuration of a membrane module. Some modules have hollow fibers opening at both ends, whereas others may have only one end open to the filtrate side, as illustrated in Figure 3-4. For the sake of simplicity, the following discussions are based on modules having one open end. For modules with both ends open, the same principle can be applied for the calculation, but bypass flow needs to count the sum of the flows from both ends through the membrane breach.

Outside-in, one filtrate side

Outside-in, two filtrate sides

Inside-out, cross-flow

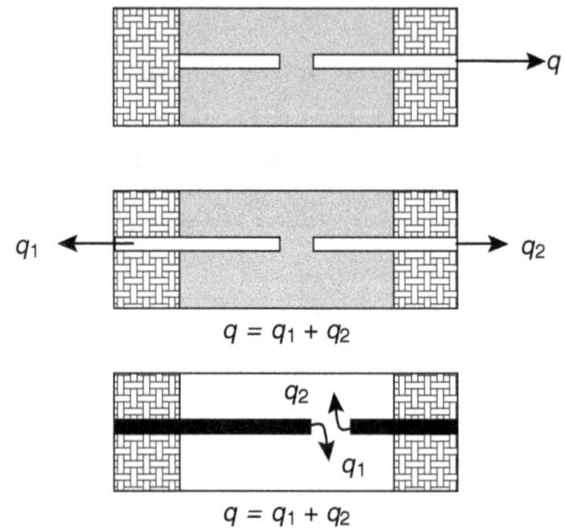

(q = total bypass flow; q_1 = bypass flow from an end; q_2 = bypass flow from the other end)

Figure 3-4. Module configuration and bypass flow

Table 3-3. Hydraulic valculation of hollow-fiber bypass flow (see appendix 3-A for the derivations of equations)

Parameter	Re ≤ 2,000	3,000 < Re < 100,000
Friction coefficient (f)	$\frac{64}{Re}$ Eq 3-12	$\frac{0.316}{Re^{0.25}}$ Eq 3-16
Major loss (γh_f)	$\frac{32\,\mu V}{D^2}$ Eq 3-13	$0.158\frac{\rho^{0.75}\,\mu^{0.25}\,V^{1.75}}{D^{1.25}}$ Eq 3-17
Flow velocity (V)	$\frac{(p_{in}-p_{out}-\gamma L)D^2}{32\mu L}$ Eq 3-14	$2.87\left(\frac{p_{in}-p_{out}-\gamma L}{L}\right)^{0.571}\frac{D^{0.714}}{\rho^{0.429}\,\mu^{0.143}}$ Eq 3-18
Bypass flow (q)	$\frac{\pi(p_{in}-p_{out}-\gamma L)D^4}{128\mu L}$ Eq 3-15	$0.718\pi\left(\frac{p_{in}-p_{out}-\gamma L}{L}\right)^{0.571}\frac{D^{2.714}}{\rho^{0.429}\,\mu^{0.143}}$ Eq 3-19

Re = Reynolds number

f = friction coefficient

V = flow velocity in fiber lumen

q = bypass flow through the fiber

D = diameter of fiber lumen

ρ = density of water

$p_{in}-p_{out}$ = pressure differential

h_f = hydraulic loss in the fiber lumen

μ = dynamic viscosity of water

γ = specific weight of water

L = length of flow path in fiber lumen

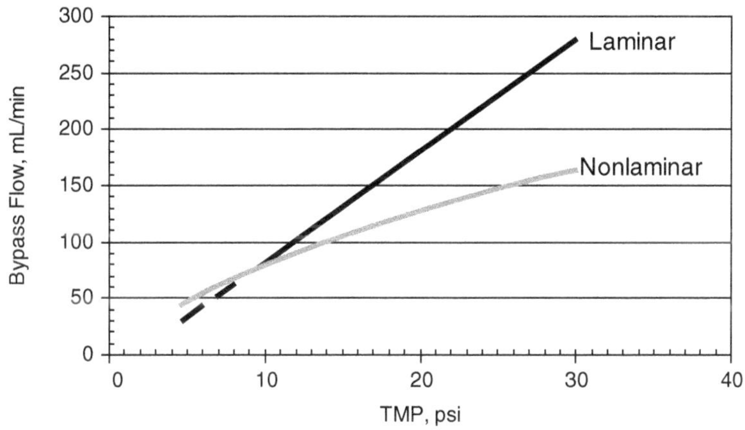

Figure 3-5. Bypass flow through a breached hollow-fiber membrane calculated using laminar and nonlaminar flow regimes (TMP = trans-membrane pressure, in pounds per square inch [psi])

The bypass flow through a single completely broken fiber is a function of pressure differential, the length of the flow path in the fiber lumen, and the diameter of the fiber lumen. This bypass flow can be derived through the principles of hydraulics such as mass and energy conservation—namely, the Bernoulli Law. If the minor losses at the entrance and exit of the hollow fiber are negligible compared with the major loss in the fiber lumen and the fiber lumen is hydraulically smooth, then Equations 3-12 through 3-19 in Table 3-3 can be employed for hydraulic calculations of bypass flow in an outside-in mode for two ranges of values of Reynolds numbers ($Re \leq 2{,}000$ and $3{,}000 < Re < 100{,}000$). The mathematical derivation of equations is given in appendix 3-A. For hydraulically rough fiber lumens (i.e., relative roughness $e/D > 0.02$ approximately), the value friction factor used in the hydraulic calculation is primarily a function of relative roughness and does not vary significantly with Reynolds number. Therefore, this value can be considered to be a constant for a given hollow-fiber lumen.

For each calculation, the value of the Reynolds number should be checked to ensure the equations used for the calculation correspond to the right flow regime. In reality, a laminar regime ($Re \leq 2{,}000$) typically occurs for a long flow path in a fine fiber lumen and a small pressure differential. A very short flow path and a large pressure differential generally correspond to a high Reynolds number.

Figure 3-5 depicts an example of the bypass flows calculated according to the equations for the different ranges of Reynolds numbers in Table 3-3. As indicated in Figure 3-5, the calculated bypass flow for laminar flow condition is larger than that for nonlaminar flow condition. This difference in bypass flow calculated from two sets of equations becomes larger as the pressure differential increases. As a result, the use of equations for laminar flow would have a more significant impact when assessing the

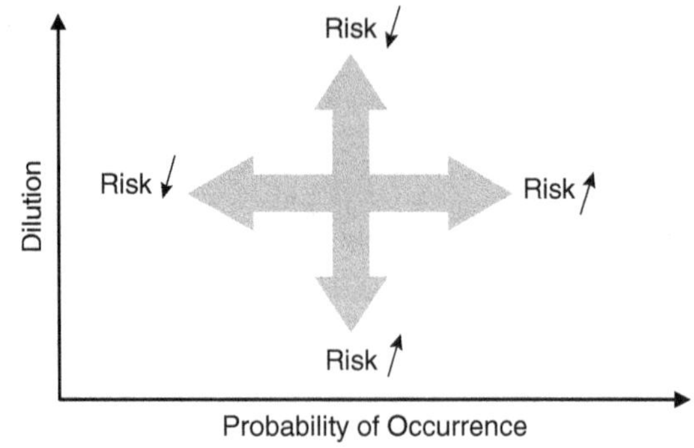

Figure 3-6. Risk assessment diagram for membrane integrity breach

effect of broken fibers on product water quality. At a high Reynolds number, the bypass flow as a function of pressure differential is more likely to follow the trend calculated with equations for a nonlaminar flow regime. In this sense, the use of laminar flow regime in the calculation of bypass flow provides a conservative estimate for the impact of fiber breakage.

Calculate LRV to assess the consequence of a membrane breach. Once the bypass flow is known, Equation 3-9 can be used to assess the consequence of a membrane integrity breach for a given membrane system for which the values of Q_P and VCF are known. The lower the LRV, the more significant the consequence of the membrane breach is. An example of LRV calculation using measured bypass flow values of a single broken hollow fiber under various conditions is presented in chapter 5.

Risk Assessment of Membrane Integrity Breach

Because risk is the product of the probability of occurrence and the consequences of a membrane integrity breach, a 2-D risk assessment diagram for membrane integrity breach can be formulated based on the occurrence and consequence (dilution) as depicted in Figure 3-6.

This diagram divides the plane into different risk zones. The upper left corner of the diagram represents a low-risk zone while the lower right corner represents a high-risk zone. Based on acceptable risks, the boundary can be drawn and a required testing frequency can be determined accordingly. It should be noted that the probability of occurrence is not only system-specific but also time-dependent. For example, at the very beginning of the system startup or near the end of filter life, the probability of occurrence can be significantly higher than at the other stages.

As the size of the membrane system increases, the probability of fiber breakage increases because the total number of fibers in the system increases. However, LRV also increases due to the dilution effect of a large system. Therefore, as the size of the membrane system increases, the filter failure (with a single defective fiber) gradually becomes a moderate- to high-probability event but with a less significant consequence caused by a low concentration of pathogens in the finished water as a result of increased dilution of bypass flow.

As the probability of occurrence of membrane integrity breach varies during the course of membrane service time, a stratified testing scheme that has more frequent testing following the startup and near the end of membrane's life cycle and less frequent testing the rest of the time is also feasible. This is because the majority of breaches from either manufacturing or improper installation would be discovered early after commissioning the system, while the membrane breaches from the degradation in strength would be discovered near the end of the membrane's life cycle. Again, this stratified approach based on the service time of membranes should recognize the effect of system size on the testing frequency because the consequences of an integrity breach differ depending on system size.

RELIABILITY

Reliability of membrane integrity testing is defined as the ability of an integrity-testing method to exclude false-positive or false-negative signals (Liu et al. 1999; AwwaRF 2000; Liu and Schaefer 2002). A false-positive signal indicates an integrity breach when no integrity breach is present; and a false negative indicates that there is no integrity breach when one is actually present. In scientific literature, reliability of a measurement sometimes refers to specificity, which implies the ability to distinguish a true response from interferences. The issue of reliability is not emphasized in LT2ESWTR, but that does not mean reliability is less critical. On the one hand, a false negative can potentially compromise the safety of drinking water. On the other hand, although it does not directly affect drinking water quality, a false positive can be annoying to operators and disruptive to plant operation. In the worst case, repeated false-positive signals could lead the operators to ignore the test results completely and miss the true positive signals.

From the statistical point of view, every measurement is a random event. Therefore, the observed difference in responses can be caused by chance—random noise that is not related to the causes of the difference in responses. An example of a random nature of particle counts in membrane filtrate is presented in Figure 3-7 in which particle counts are plotted at varying degrees of membrane breach. Although a trend of increasing counts can be observed as membrane breach became worse (more broken hollow fibers), counts were overlapped so much that a single measurement may not be indicative of whether the membrane is integral (e.g., considering counts at points A and B in the figure). This random component of the measurement forms the background noises of measurements.

Unless a response is sufficiently strong rising above the background noises, the determination of its significance can be very difficult, if not impossible. When the level of background noise is high relative to the "true" response, the judgment of

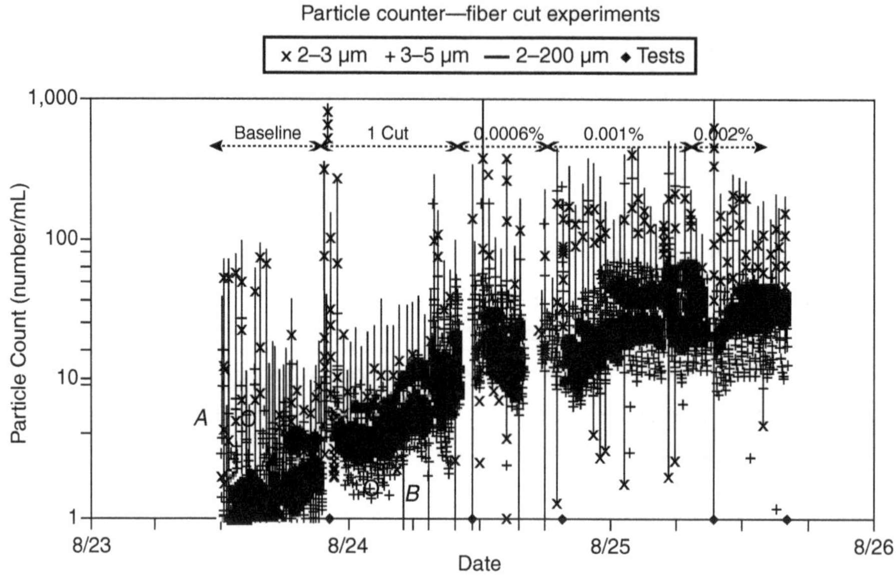

Figure 3-7. Particle count of varying degrees of membrane breaches shows a random nature of measurement (courtesy of Sethi et al. 2004)

the significance of any response becomes a matter of whether a particular response is meaningful (e.g., signifying a membrane integrity breach).

As the purpose of integrity testing is to verify that a membrane system is removing the target contaminant as it should, a test must be able to confirm whether the objectives are achieved. In other words, integrity testing needs to set certain control limits (CLs) above which certain actions need to be taken to ensure that water quality criteria are met (USEPA 2005). There are typically two levels of control limits: lower control limit (LCL) and upper control limit (UCL). An LCL is the level at which the operator is alerted that an integrity breach may have occurred, but it has not yet reached the point of compromising water quality objectives. A UCL is the level at which corrective actions must be taken, or the water quality objectives would be in jeopardy.

Many integrity verifications based on monitoring filtrate quality are particularly subject to the influence of the random nature of the measurement (see chapter 6). Therefore, it is essential to know certain statistical characteristics of the sample population in those measurements in order to establish the UCL. If the sample distribution follows Gaussian distribution (the bell curve), statistical parameters such as mean and standard deviation of the samples can be used to establish a control limit.

Figure 3-8 is a graphic representation of such an approach. The scale factor Z in Figure 3-8 is to measure the difference between a particular sample and the mean of the sample population. In other words, the larger the difference, the lower the probability

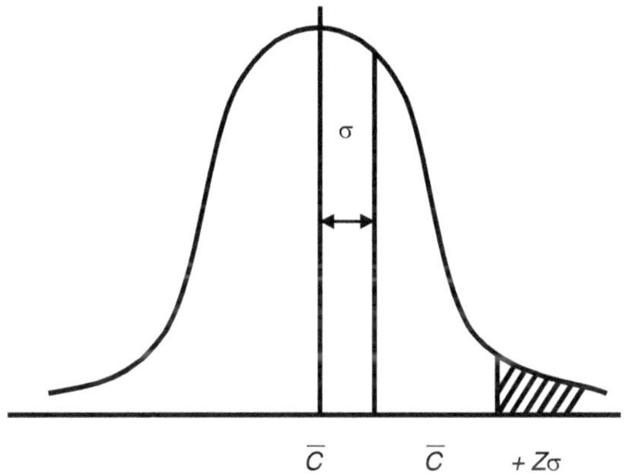

Figure 3-8. Gaussian distribution and population statistical description ($\bar{C}$ = mean; σ = standard deviation; *Z* = scale factor)

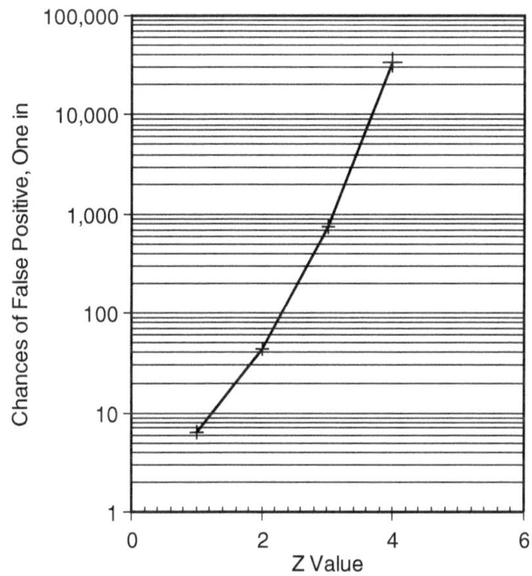

Figure 3-9. Chances of false positive as a function of *Z* value

that this sample belongs to the sample population. Illustrated graphically, for a given sample population described by its mean $\bar{C}$ and standard deviation σ, the probability of samples with value greater than $\bar{C} + Z\sigma$ belonging to the same population is the shadowed area at the right side in Figure 3-8. If $\bar{C}$ and σ share the mean and standard deviation of baseline measurements of an integral membrane system, respectively, a UCL can be established based on a value of Z to have a known probability of false positives. The relationship of probability of false positives and Z value is depicted in Figure 3-9. The probability of false positive for $Z = 3$ is one in 740 measurements, or 99.86 percent. This seemingly fairly low percentage of false-positive signals can still

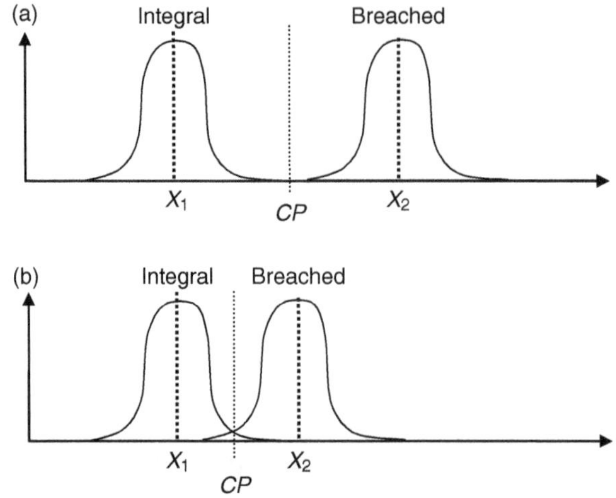

Figure 3-10. Selecting the control point (CP)

generate a significant number of false alarms if the number of measurements gets very large. For example, if a particle counter reads once every 15 seconds and a membrane system is in production 90 percent of the time on daily average, the counter would generate about 7 false alarms every day when the UCL for the system is set at $Z = 3$.

When a membrane integrity breach cannot be reliably detected, the setting of UCL becomes a balancing act between avoiding false-positive and false-negative signals. Figure 3-10 illustrates the situation. In Figure 3-10(a), there is no overlap between the distribution curves for the integral and the breached membranes. Selecting UCL is straightforward. In Figure 3-10(b), there is an overlap of two curves. Therefore, UCL would likely produce false-positive signals, false-negative signals, or both, depending on where the vertical line representing UCL is drawn. The confidence level on which a measurement is judged to indicate system integrity depends on the degree of the overlap of two distribution curves. Therefore, reliability of an integrity method also can be defined by the degree of overlapping of distributions of measurements for integral and breached situations.

In an integrity-testing method, sensitivity and reliability are usually related. A method with low sensitivity cannot reliably detect a membrane breach that is close to the method's detection limit. Conversely, a method with low reliability cannot be deemed sensitive because of the uncertainty of the result.

Although reliability of integrity testing is discussed in general terms, it is the most relevant to the methods based on monitoring filtrate quality. This point is illustrated by many examples in chapter 6 in which the importance of reliable measurement for verification of membrane integrity is highlighted.

TRACEABILITY

Traceability is defined as the ability of a testing method to locate where the breach occurs within a membrane system. Like reliability, traceability is not a regulatory issue and receives very little attention. By no means, though, is traceability less important for membrane plant operators; it is a fundamental part of a comprehensive integrity testing procedure. As was discussed in chapter 2, a low-pressure membrane plant is modular in configuration: it typically consists of multiple skids and each skid consists of multiple filter modules. A filter module usually consists of thousands hollow fibers or more. Therefore, traceability means not only locating and identifying the compromised module(s) but also locating the breach from a particular hollow fiber or fibers among the many. For a typical membrane plant with a capacity of a few million gallons per day, locating a particular membrane breach is like trying to find a needle in a haystack.

The key measure of traceability is how quickly a breach can be located and isolated. As the procedure of locating the breach is typically performed when the system is off-line, the time spent to locate the breach means a loss in productivity.

Tracing the membrane integrity breach requires ways to identify the breach and means to isolate the breached module(s) from the rest of the system. Those features have to be considered in the design and engineering of the membrane system. Another important design consideration is easy access to the modules so that isolation and repair and/or replacement of modules can be done easily and quickly.

IMPLEMENTABILITY

Implementability refers to the easiness to implement a membrane integrity monitoring method. It includes the installation of testing /monitoring hardware and software, operating and maintenance associated with membrane integrity testing, and operator skills and training required to perform those tasks. In the MFGM, it is suggested that utilities establish a comprehensive integrity verification program. The reader can refer to Appendix A of the MFGM for details of the program. Implementability of integrity methods can help or hinder the execution of such a program. Some examples will be presented in chapter 8.

SUMMARY

In this chapter, important issues for membrane integrity testing have been discussed. These issues include resolution, sensitivity, testing frequency, reliability, traceability, and implementability—common threads, regardless of the integrity method employed. When assessing an integrity-testing method or comparing different methods, these issues can be used to form a matrix for evaluation. The discussion of particular methods for integrity testing in the following chapters will focus on how each method measures up in terms of resolution, sensitivity, testing frequency, reliability, and traceability.

Appendix 3-A: Derivation of Equations 3-12 Through 3-19

The diagram for mathematical derivation is given in Figure 3-A-1. The orientation of the hollow fiber is vertical and the fluid flows upward.

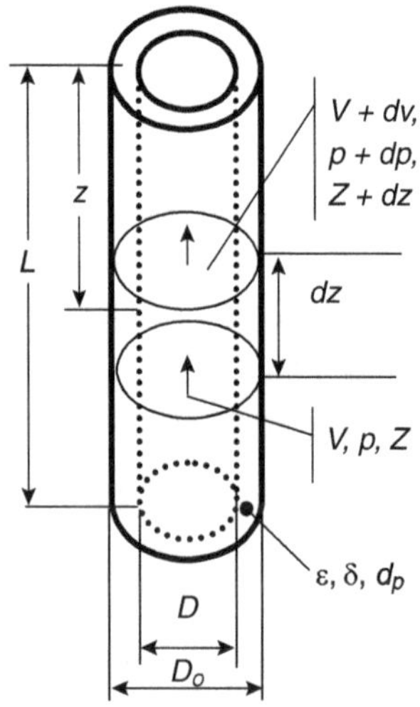

Symbols:

$V, p, Z =$ flow velocity and pressure inside the hollow fiber, and elevation at the bottom of the fiber segment, respectively

$V + dv, p + dp, Z + dz =$ flow velocity and pressure inside the hollow fiber and elevation, and elevation at the top of the fiber segment, respectively

dz = height of the segment of hollow fiber

u = flux of flow across the membrane in the segment

D, D_O = inside and outside diameter of the hollow fiber, respectively

L = overall fiber length

ε = the porosity of the hollow-fiber membrane

δ = wall thickness of the hollow-fiber membrane

d_P = the mean pore size of the hollow-fiber membrane

Figure 3-A-1. Diagram for derivation of equations 3-6 through 3-13

Applying the Bernoulli Equation to the hollow-fiber segment:

$$Z + \frac{p}{\gamma} + \frac{V^2}{2g} = Z + dZ + \frac{p + dp}{\gamma} + \frac{(V + dV)^2}{2g} + h_f \qquad \text{(Eq 3-A-1)}$$

Where

γ = specific weight of water, kN/m^3

h_f = hydraulic loss in the fiber segment, m

Rearrange the equation and neglect the small term including dV^2:

$$dZ + \frac{dp}{\gamma} = -\frac{VdV}{g} - h_f \qquad \text{(Eq 3-A-2)}$$

Or:

$$\frac{dp}{dZ} = -\gamma - \frac{\gamma VdV}{g} - \gamma h_f \qquad \text{(Eq 3-A-3)}$$

The last term on the right-hand side of Equation 3-A-3 represents the hydraulic loss of the segment, which can be described by Darcy's Equation:

$$h_f = \frac{f \cdot dz \cdot V^2}{2Dg} \qquad \text{(Eq 3-A-4)}$$

Where

f = Darcy's friction coefficient

If the flow in the hollow-fiber lumen is laminar flow, then

$$f = \frac{64}{\text{Re}} = \frac{64\mu}{\rho DV} \qquad \text{(Eq 3-A-5)}$$

Where

Re = Reynolds number

If the value of the Reynolds number is in the range of 3,000 to 100,000, then Blasius Equation can be used for calculate the friction coefficient (Giles et al. 1994):

$$f = \frac{0.316}{\text{Re}^{0.25}} \qquad \text{(Eq 3-A-6)}$$

Now consider the case of laminar flow in the fiber lumen. Substitute Equation 3-A-4 into Equation 3-A-3:

$$h_f = \frac{32\mu V dZ}{\rho g D^2} = \frac{32\mu V dZ}{\gamma D^2} \qquad \text{(Eq 3-A-7)}$$

Substitute Equations 3-A-4 and 3-A-7 into Equation 3-A-3 and rearrange:

$$\frac{d_p}{dZ} = -\gamma - \rho V \frac{dV}{dZ} - \frac{32\mu V}{D^2} \qquad \text{(Eq 3-A-8)}$$

According to the mass balance for the flow through the membrane of the segment, the incremental increase in flow in the segment is equal to the flow infiltrated from the segment, i.e.,

$$\frac{\pi}{4} D^2 dV = \pi D_o u dZ$$

Or:

$$\frac{dV}{dZ} = \frac{4D_o}{D^2} u \qquad \text{(Eq 3-A-9)}$$

Membrane flux, u, can be calculated via Hagen-Poiseuille Equation:

$$u = \frac{\varepsilon\, dp^2\, \Delta P}{32\mu\sigma} \qquad \text{(Eq 3-A-10)}$$

Where

ΔP = pressure differential across the membrane, or trans-membrane pressure.

Substitute Equation 3-A-10 into Equation 3-A-9:

$$\frac{dV}{dZ} = \frac{D_o \varepsilon\, dp^2\, \Delta P}{8D^2\, \mu\sigma} \qquad \text{(Eq 3-A-11)}$$

Substitute Equation 3-A-11 Equation into Equation 3-A-8:

$$\frac{dp}{dZ} = -\gamma - \rho V \frac{D_o \varepsilon \, dp^2 \, \Delta P}{8 D \mu \sigma} - \frac{32 \mu V}{D^2} \qquad \text{(Eq 3-A-12)}$$

On the right-hand side of Equation 3-A-12, the first term represents the contribution of elevation (i.e., gravity) to the spatial change in pressure inside the fiber lumen, or static hydraulic head; the second term represents the contribution of infiltration through the membrane to the spatial change in pressure inside the fiber lumen; and the third term represents the contribution of hydraulic loss to the spatial change in pressure inside the fiber lumen. For a fine membrane where pore size, *dp*, is very small, the second term is almost two orders of magnitude smaller than the first and third terms. Therefore, Equation 3-A-10 can be simplified as

$$\frac{dp}{dZ} \approx -\gamma - \frac{32 \mu V}{D^2} \qquad \text{(Eq 3-A-13)}$$

In essence, Equation 3-A-13 says that for a fine membrane, the effect of infiltration across the membrane on fluid dynamics inside the fiber lumen of a compromised fiber can be practically neglected. That is, the flow velocity inside the fiber lumen, *V*, can be considered as a constant along the flow path.

Rearrange and take integral of Equation 3-A-13:

$$\int_0^L dp = -\left(\gamma + \frac{32 \mu V}{D^2}\right) \int_0^L dZ$$

Or:

$$p_{in} - p_{out} = \left(\gamma + \frac{32 \mu V}{D^2}\right) L \qquad \text{(Eq 3-A-14)}$$

Where

p_{in} and p_{out} = the pressure at the inlet and outlet of hollow fiber, respectively

Taking minor losses (hydraulic losses from fiber inlet and outlet) into account:

$$p_{in} - p_{out} = \left(\gamma + \frac{32 \mu V}{D^2}\right) L + \frac{(C_1 + C_2)}{2g} V^2 \qquad \text{(Eq 3-A-15)}$$

Where

C_1 and C_2 = loss coefficients for fiber inlet and fiber outlet, respectively

Using a similar approach, one also can derive the equation similar to Equation 3-A-15 for nonlaminar flow where Reynolds number falls in the range of 3,000 to 100,000:

$$p_{in} - p_{out} = \left(\gamma + 0.158 \frac{\rho^{0.25} \mu^{0.75} V^{1.75}}{D^{1.25}}\right) L + \frac{(C_1 + C_2)}{2g} V^2 \qquad \text{(Eq 3-A-16)}$$

Where

ρ = the density of water, kg/m^3

Typically, minor losses from entrance and exit of a hollow fiber are considerably less likely than major loss. That is, the last terms in Equations 3-A-15 and 3-A-16 can be omitted:

$$p_{in} - p_{out} = \left(\gamma + \frac{32\mu V}{D^2}\right)L \qquad \text{(Eq 3-A-17)}$$

$$p_{in} - p_{out} = \left(\gamma + 0.158\,\frac{\rho^{0.25}\,\mu^{0.75}\,V^{1.75}}{D^{1.25}}\right)L \qquad \text{(Eq 3-A-18)}$$

From Equations 3-A-17 and 3-A-18, the bypass flow can be calculated by multiplying the flow velocity, V, and the area of the cross-section of fiber lumen, $\pi D^2/4$. Major equations for hydraulic calculations are summarized in Table 3-3.

4

TYPES OF MEMBRANE INTEGRITY TESTING

INTRODUCTION

In chapter 3, the terms *direct* and *indirect integrity testing* are used in the discussion of regulatory requirements for membrane systems in LT2ESWTR. According to the definition in LT2ESWTR, direct integrity testing is "a physical test applied to a membrane unit in order to identify and isolate integrity breaches"; and continuous indirect integrity monitoring is defined as "monitoring some aspect of filtrate water quality that is indicative of the removal of particle matter" (USEPA 2006a). The *Membrane Filtration Guidance Manual* deliberates further on direct integrity testing: "There are two general classes of direct integrity tests that are commonly used in membrane filtration facilities: pressure-based tests and marker-based tests."

According to this statement, the key distinction between direct and indirect integrity testing is essentially to emphasize the sensitivity of the testing. In short, direct integrity testing is that with high sensitivity to signify a membrane integrity breach, while indirect integrity testing lacks such sensitivity. This makes sense from a regulatory perspective: the ultimate goal for membrane integrity testing is to safeguard drinking water quality.

In this book, a slightly different classification system is used. The physical processes for methods of testing membrane integrity current in the industry can be generally divided into two categories: One is based on measurements of fluid flow or movement; the other is based on measurements of filtrate quality. The first category includes methods such as pressure-decay, vacuum decay, diffusion flow, water replacement, and so on. Regardless of the name, the physical process is the same—fluid flow and movement under a pressure differential. The differences are the condition (pressure versus vacuum), the measured parameters (pressure or flow), and fluids (liquid versus gas). The second category includes any method for measuring filtrate quality—such as turbidity monitoring, particle count, and other measurement methods, regardless of whether the measurements are for indigenous items or artificially spiked markers.

Therefore, the first category in this book includes all direct integrity testing methods except the marker-based test, which is discussed in the second category. The marker-based test uses the same process as other methods based on monitoring filtrate quality but with injection of a specific marker upstream of membranes. The first and the second categories are discussed in chapters 5 and 6, respectively. In addition, there are integrity-testing methods that do not belong to either category. Examples of such methods are the acoustic method using hydrophones, particle amplifier, and relative trans-membrane monitoring, which are discussed in chapter 7.

INTEGRITY TESTING BASED ON FLUID FLOW

The basic principle of this class of integrity testing is that the capillary force in a membrane pore is related to the diameter of the pore. When air pressure is applied on one side of a liquid-filled membrane and air pressure overcomes the capillary force in the pore, the liquid in the membrane pore is pushed out. Once the liquid is pushed out of the pore, air flows through the pore with little resistance (not unlike a deflated tire that is punctured by a nail). The air pressure just above the capillary force in the liquid-filled membrane pore is called the *bubble point*, referring to the fact that a stream of air bubbles occurs as the result of air flowing through the pore.

Figure 4-1 illustrates the bubble-point concept. As indicated in Figure 4-1, if the pressure applied is less than the bubble point (P_B), air passes through membrane pores via diffusion only. Conversely, if the pressure applied is above the bubble point, a steady stream of bubbles appears on the side of liquid-filled membrane. Note that an actual membrane may have a range of pore sizes. Therefore, bubble point would represent the air breakthrough of the largest pores. To apply this principle to integrity testing, the pressure employed has to be less than the bubble point of the membrane, but it should be sufficiently high to enable detection of a defect with the size of the targeted contaminants, or meeting the resolution criterion as defined in chapter 3.

The relationship between air pressure and pore diameter can be described mathematically through Cantor's Equation:

$$\Delta P = \frac{4\sigma\cos\theta}{D} \qquad \text{(Eq 4-1)}$$

Where

D = diameter of defect (assuming it is round)

σ = surface tension of water

θ = water contact angle of membrane medium, characterizing membrane hydrophobicity

ΔP = bubble point—the pressure differential between upstream and downstream of the membrane

Cantor's Equation indicates the equilibrium of forces at the liquid/gas interface in a capillary vessel such as membrane pores, as illustrated in Figure 4-2.

As illustrated in Figure 4-2, the force in the gas phase projected horizontally is the product of pressure differential, ΔP, and the projected area, $\pi D^2/4$. In the liquid phase, the force projected horizontally is the product of the surface tension projected

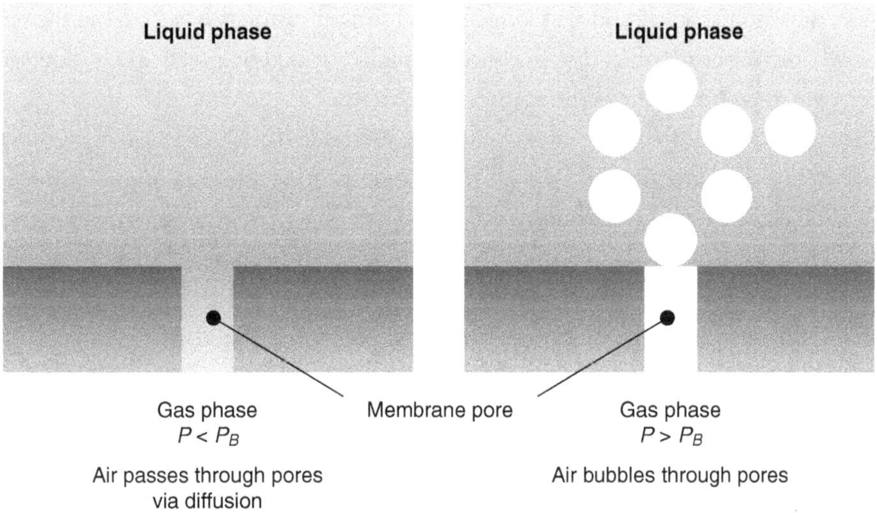

Figure 4-1. Bubble-point concept

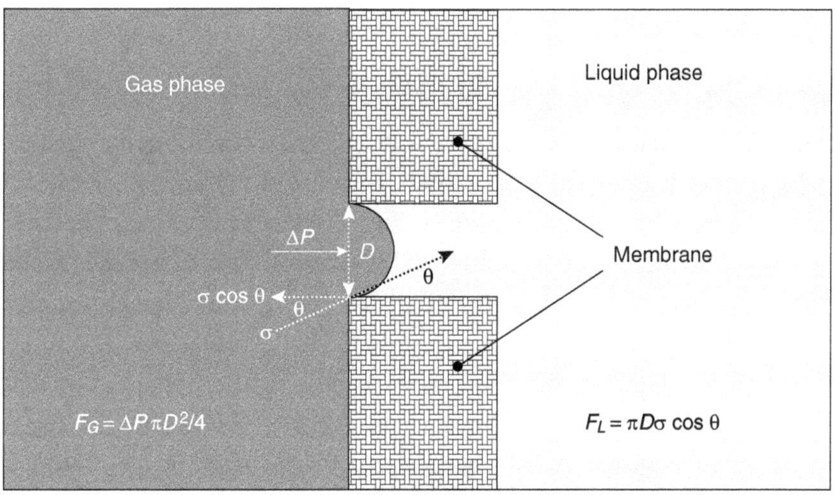

Figure 4-2. Schematic diagram for derivation of Cantor's Equation

horizontally, σcosθ, and perimeter of the pore, πD. When the forces in the interface are in equilibrium, the equation is:

$$\Delta P \times \pi \frac{D^2}{4} = \sigma \times \cos\theta \times \pi \times D \qquad \text{(Eq 4-2)}$$

The rearrangement of Equation 4-2 is Cantor's Equation (Equation 4-1).

The concept of resolution was discussed in chapter 3. Resolution is the minimum size of a defect that is detectable by integrity testing. Because the pore size of a given membrane can be related to its bubble point as indicated by Cantor's Equation, one can theoretically select a testing pressure for a targeted contaminant with known size to meet the resolution criterion, as long as the testing pressure is less than bubble point. However, the selection of testing pressure is usually limited by practical considerations such as the pressure rating of the membrane systems.

There are several forms of integrity testing methods. In ASTM D-6908-03 (Standard Practice for Integrity Testing of Water Filtration Membrane Systems), pressure decay and vacuum decay methods are listed. Table 4-1 presents four methods, representing the spectrum of the methods in this category. It becomes clear that in spite of measuring different parameters, the underlying processes are the same for all of the methods: air flows and/or water flows under the pressure differential.

It should be noted that in practice, the integrity-testing methods in this category detect not only the defects of membranes but also any air leakage in the part of the system that is under the air pressure, including values and joints of piping. Therefore, air tightness of the part of the membrane system that is under air pressure during the testing is very important for this type of integrity testing. The impact of this factor will be discussed in chapter 8.

INTEGRITY TESTING BASED ON MEASURING FILTRATE QUALITY

The practice of evaluating the treatment system's efficiency by monitoring effluent quality has a history almost as long as water treatment itself and is considered to be the benchmark in drinking water treatment. In current practice for low-pressure membrane systems, the monitoring of water quality is typically accomplished via on-line monitoring by instruments such as the turbidimeter or particle counters. These give the operator nearly real-time information regarding the performance of the system.

At first glance, methods based on water quality monitoring are intuitive: when the membrane integrity breaches, filtrate quality declines. However, on closer examination, many issues emerge. A few examples are: What should be monitored? Where should be sampled? How sensitive and reliable are the results? Many of these issues are based on the fact that the filtrate flow from intact membranes is typically much larger than the bypass flow from the membrane breach. As the bypass flow is blended with the filtrate flow, the sensitivity and reliability of the measurement are lowered by the dilution effect.

Recall the description of a typical membrane plant of hollow-fiber configuration. There are thousands and thousands individual hollow fibers in a membrane module; and a membrane skid or rack contains multiple modules. Therefore, it is easy to conclude that the contribution to overall water quality from the bypass flow of a small membrane defect practically is insignificant, even negligible. Monitoring techniques (e.g., turbidity and particle count) for conventional media filters may not register the small change in filtrate quality caused by the membrane defect.

Table 4-1. Different integrity methods based on fluid flow

Method	Schematic	Description	Measured Parameter
Pressure decay	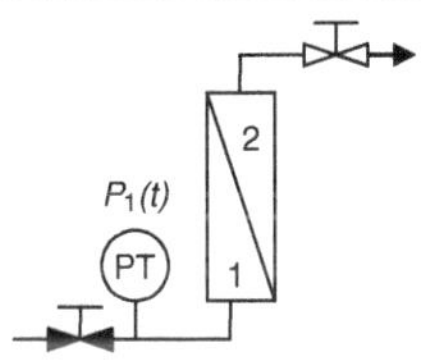	One side of the membrane (1) is pressurized with air and the valve is closed. On the other side of the membrane (2) the valve is open to allow the passage of air. Side 2 can be filled with water or air.	Pressure decay over time $P_1(t)$.
Vacuum decay	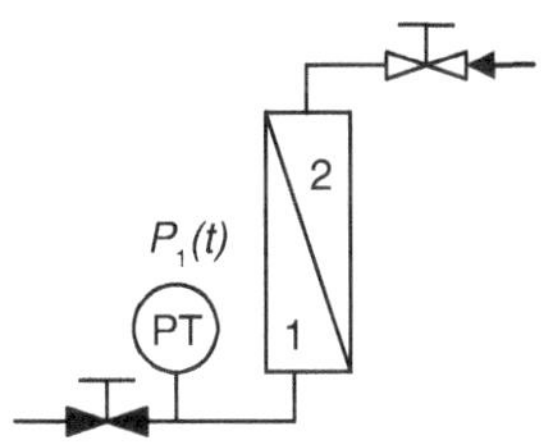	Similar to pressure decay, but Side 1 has negative pressure (vacuum) and Side 1 is typically filled with air.	Vacuum decay over time $P_1(t)$.
Diffusion flow	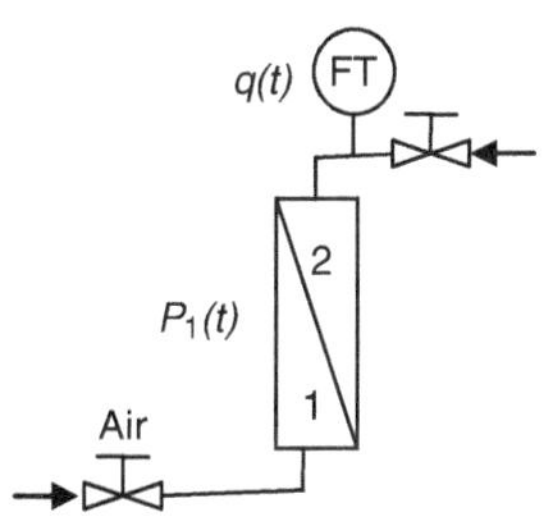	Continuous air supply on one side of the membrane (1). Valves on both sides of the membrane are open to allow air passage.	Airflow rate $q(t)$.
Water replacement	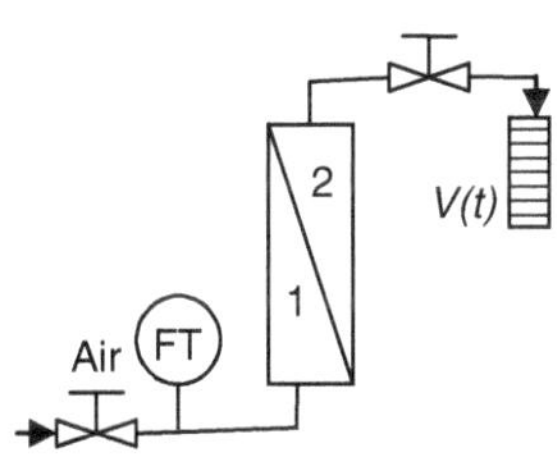	Continuous air supply on one side of the membrane (1). Side 2 is filled with water. Valves on both sides of the membrane are open to allow air passage.	Volume of replaced water by air over time $V(t)$.

There are several ways to improve the sensitivity of integrity-testing methods based on monitoring the filtrate quality, including the following:

- *Improve the sensitivity of instrumentation.* For example, using a laser turbidimeter increases the sensitivity of the measurement by one to two orders of the magnitude (Sadar 2004). Also, the use of a high-resolution particle counter improves sensitivity (Panglisch et al. 1998; Kruithof, Kamp, and Holmer 2001; Sethi et al. 2004).
- *Use multiple sensors.* This approach improves sensitivity as each sensor would measure a portion of filtrate and consequently reduce the dilution effect

(Banerjee, Lozier, and Carlson 2001; Sadar 2002b; Sadar et al. 2003a). Alternatively, membrane skid or rack may be divided into two or more segments isolated by valves to reduce the dilution effect and/or restrain volume to enhance testing sensitivity (Vickers 2009).

- *Improve sensitivity by modifying the way the data are processed* (Banerjee, Lozier, and Carlson 2001). The basic idea of this approach is that even though the average reading of the filtrate quality indicator is low because of the dilution effect, the variations in reading do increase as the instrument registers the change at a certain time. As a result, even though the difference in average of readings of a system with a membrane defect and one without is statistically insignificant, the system with a membrane defect would show a greater variation in readings.
- *Use feed spike to enhance the sensitivity* (Franklin, Smith, and Knops 2000; Franklin, Smith, and Knops 2001; Van Hoof, Kruithof, and Kamp 2001). This approach is deemed the marker-based method in the MFGM (USEPA 2005). The spiking reagents are similar to indigenous substances in the source water or to reagent used in water treatment plants (e.g., powdered activated carbon particles). The spiking reagents are basically surrogates for water quality. By extension, foreign substances can also be used if they make detection more sensitive. However, the use of spiking reagents that are not indigenous can be subject to regulatory scrutiny to ensure that they do not pose safety and environmental concerns themselves. One possible solution is to use spiking reagents that are already certified to be safe for drinking water by the National Sanitation Foundation (NSF), even though the perception of health and safety from injecting foreign substances into water by the customers may still need to be addressed.
- *Use a contaminator concentrator.* The concept is based on the fact that the filtrate quality in intact low-pressure membranes is generally high, and consequently filtrate concentration is required to enhance detection sensitivity. This concept was applied in a study in which a small backwashable filter was placed on the filtrate line to concentrate particles in the filtrate. Then the filter was backwashed and the particles in backwash water were counted in a batch mode by a particle counter (Sethi et al. 2004). Depending on the duration necessary for filtrate concentration, this method may not be considered as continuous monitoring. At the current form, the method may not be practical for utilities as it requires significant involvement of the plant operator. Great improvement to the method and automation are necessary for implementation of the method.

COMPARISON OF INTEGRITY METHODS

A simple comparison of two categories of methods is presented in Table 4-2. The comparison is based on the relative capacities for meeting critical criteria: i.e., resolution,

Table 4-2. Comparison of integrity testing methods

Testing Method	Resolution	Sensitivity	Reliability	Continuity	Traceability	Implementability
Based on fluid flow	Depending on testing pressure, generally capable of a few micrometers	Depending on severity of breach, testing pressure, and air tightness of the system, typically greater than 3-logs	Depending on severity of breach, testing pressure, and air tightness of the system, generally reliable	A batch test to be performed only when the unit is off-line	Capable of tracing to the individual filter module with clear housing or fittings	Easy to implement
Based on water quality monitoring	Depending on the monitoring instrument, ranging from micrometers to dissolved species	Depending on source water quality and sensitivity of the instrument used, generally not sensitive to small breaches*	Depending on source water quality and the instrument used, generally not reliable	Depending on the sample processing time, generally capable of measuring in an interval of a few seconds when system is running†	Capable of tracing to the individual filter module only in multiple-sensor arrangement	Easy to implement, but maintenance of the monitoring instrument is a must

*With the exception of marker-based (feed spike) test, which can have high sensitivity.
†With the exception of contaminant concentrator, which is typically noncontinuous. Marker-based test typically is not continuous for practical reasons.

sensitivity, continuity, reliability, traceability, and implementability. Overall, methods based on fluid flow are more sensitive and reliable, but the tests have to be performed periodically when the system is off-line. Methods based on monitoring water quality are low in sensitivity and reliability, but they can provide continuous monitoring. One exception may be the marker-based integrity testing. As a practical matter, a continuous spiking of marker into the feed stream of the membrane systems may not be practical. Therefore, a typical approach is to perform marker-based integrity testing periodically.

The pros and cons for each category of methods and the factors affecting their capacities will be discussed in detail in chapters 5 and 6, respectively.

5

MEMBRANE INTEGRITY TESTING METHODS BASED ON FLUID FLOW

INTRODUCTION

Membrane integrity testing based on fluid flow involves creating a pressure differential across the membrane by pressurizing (or creating a vacuum) on one side of membrane with a gas (typically air) and measuring the indicator of the movement of the fluid (air) from one side of the membrane to the other. This indicator can be pressure decay, vacuum decay, air flow, or the flow displaced water, depending on the particular method used. The applied pressure (vacuum) is lower than the bubble point of the membrane. Because of the surface tension of the water in membrane pores, the pressure or vacuum would not be able to push water out of pores. However, if there is a membrane breach and the applied pressure (vacuum) exceeds the bubble point, water in the membrane breach would be pushed out. As a result, air passage through the breached membrane increases considerably. By comparing the value of the parameter indicating air passage with that of the integral membranes, one can determine whether the membrane is breached and the degree of the membrane breach when it occurs. This method is illustrated in Figure 5-1.

As discussed in chapter 4, methods based on the flow of fluid come in many forms: pressure decay, vacuum decay, forward flow, and water displacement, to name a few. Regardless of the form, all those methods are based on the same working principle: flow of fluid through the membrane breach under a pressure differential during the testing. That is, all of these methods depend on the measurement to indicate the difference in fluid flow between integral and breached membranes.

Among these methods, the pressure-decay test is the most widely used. This method is described in great detail in the USEPA's *Membrane Filtration Guidance Manual*. Many state regulatory agencies specify it for low-pressure membrane systems for drinking water filtration. Although it is possible to use the vacuum-decay test, the consideration of vapor pressure at various temperature conditions needs to be included in developing equations and this makes the technique more complicated than the pressure-decay test. Therefore, the discussions in this chapter will focus on pressure-decay test only.

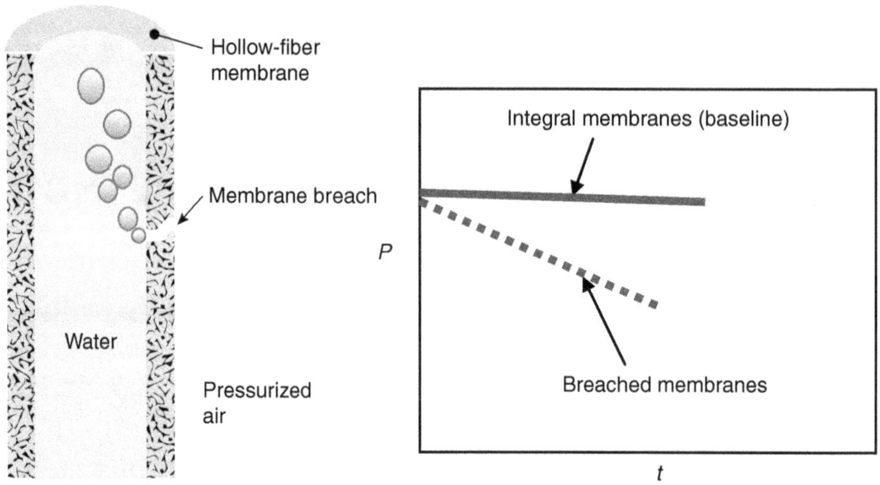

Figure 5-1. Concept of membrane integrity testing based on fluid flow (t = time; P = pressure)

Integrity testing methods based on the flow of fluid have been used for many years in the field of membrane filtration, especially in the processes where sterilized filters are required. In those applications, the result of integrity testing is correlated to tier reduction of the targeted contaminants or their surrogates by a validation test. Filter elements that pass the criterion of the integrity test are able to achieve log reduction as defined via validation. Therefore, those integrity testing methods have been a part of quality assurance/quality control (QA/QC) of production for membrane manufactures, and there are specialized integrity testing kits from different manufacturers.

There are a few unique aspects of membrane integrity testing in drinking water industry:

- Emphasis is on the integrity of the overall membrane system, not individual elements.
- The test has to be carried out in-situ.
- The size of the system tends to be much larger.
- The test needs to be performed frequently.
- It is not a merely "pass or fail" test; the results have to be quantitative (i.e., the exact value of log reduction).

These aspects define the specific requirements of membrane integrity testing in the industry. The ASTM D-6903 Practice A "Pressure Decay (PDT) and Vacuum Decay Tests (VDT)" provided the methods to determine the resolution and sensitivity in 2003. In the same year, USEPA released a draft version of the *Membrane Filtration Guidance Manual* (MFGM) and finalized the manual two years later. In this manual, key performance criteria such as resolution and sensitivity are discussed extensively.

The objective of this chapter is to answer the following questions:

1. How to determine the resolution of an integrity test using fluid-flow–based method? What are the factors affecting the determination?
2. How to determine the flow of fluid through integral and breached membranes?
3. How can parameters measured during the integrity test using fluid-flow–based methods be related to LRV? What are the factors affecting the sensitivity of the detection?
4. What are the differences among the various methods (e.g., the ASTM method, the MFGM method)?

The organization of this chapter basically follows these four questions. First, determination of resolution using fluid-flow–based method and factors affecting the determination will be discussed. Then, the discussion proceeds to the diffusion of air in water through the integral membranes and the air passage through a breached hollow-fiber membrane and how this air passage is related to the pressure decay. After that, various methods to calculate log reduction value (LRV), including those in the ASTM D6903, the MFGM, and an alternative method developed by the author, are presented. Sensitivity analyses using various equations are performed to demonstrate how various parameters in the equations affect the results of LRV calculation. Data from both microbial challenge tests and flow measurements are used for the validation of those equations. Finally, different methods for LRV calculation are compared and discussed. Due to the nature of this chapter, quite a few equations are involved. Efforts are made to present the basis and the process of developing those equations. The emphasis is on understanding of the physical concepts and basic principles of the methods. Key aspects of developing equations to describe the process will be presented; but detailed derivations for those equations will be placed in appendices to keep the content flow.

DETERMINATION OF METHOD RESOLUTION

As discussed in chapter 4, resolution refers to the minimum size of a membrane breach that can be detected. For the integrity testing based on flow, this criterion is the equivalent of applying a testing pressure above a certain value, which relates to the size of the breach through the bubble point of membrane media. This minimum size is related to the testing pressure via the concept of bubble point, which is expressed mathematically by Cantor's Equation (Equation 4-1). Cantor's Equation states that the size of a pore is inversely related to the pressure differential and varies with a membrane's water contact angle, as depicted in Figure 5-2 in which the water contact angle is equal to zero. To meet the resolution criterion of detecting a pore size of 3 μm or less as required by the LT2ESWTR, Figure 5-2 indicates that the pressure differential across the membrane has to be 14.5 psid or higher. For the testing

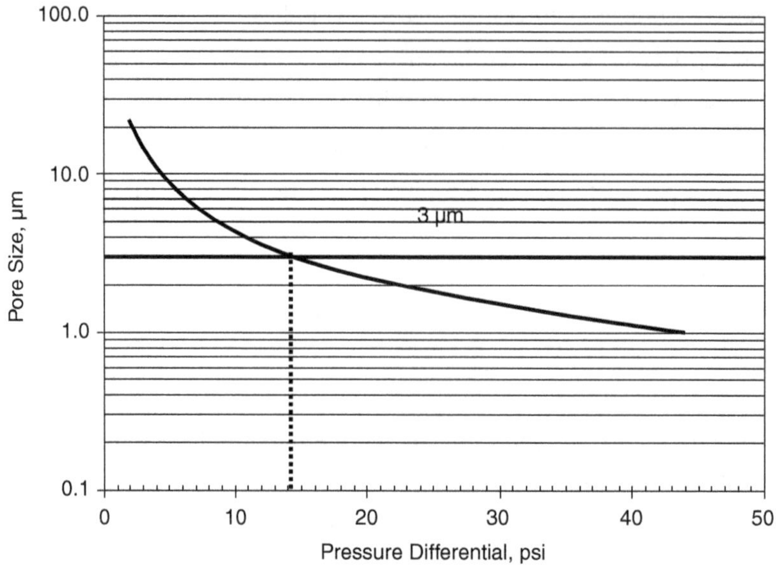

Figure 5-2. Pore size as a function of testing pressure for zero water contact angle and 32°F (µm = micrometer; psi = pounds per square inch)

pressure used in a pressure-decay test, the hydraulic head and back-pressure of the system have to be added to ensure the detection.

Pore size–pressure differential plot in Figure 5-2 describes an ideal condition where the pore is perfectly round and the water contact angle is zero. For an actual membrane, Cantor's Equation can be modified for membranes of nonround pores by introducing pore-shape correction factor, κ, as in the following;

$$D = \kappa \frac{4\sigma\cos\theta}{\Delta P} \qquad \text{(Eq 5-1)}$$

The shape-correction factor, κ, has a value between 0 and 1 where the value of 1 represents perfect round pore. The farther the pore shape is from perfect round, the smaller the value of κ. Equation 5-1 essentially implies that in nonround pores, a lower air pressure is needed to overcome the surface tension than in a round pore. The membrane pores are not "bullet holes" with straight walls; rather they are more like "an aggregation polyhedral open-walled cells of different sizes" (Meltzer 1987). As a result, the pore shape correction factor would reflect the impact of irregularity both on the surface and along the depth of the membrane medium.

Another factor affecting the testing is water contact angle, θ. The larger the value of θ, the lower the bubble point of a membrane, as implied by Equation 5-1. The water contact angle of a membrane is related to its hydrophobicity; the more hydrophobic of a membrane, the larger the value of its water contact angle.

In theory, if the precise values of κ and θ are known, the exact bubble point of a membrane can be calculated via Equation 5-1. However, this is impractical, if not impossible. There are several reasons for this:

- The contact angle term in Equation 5-1 is applicable only if the membrane surfaces are completely dry. Even if there is a single layer of water molecules on the membrane surfaces, the contact angle, θ, in the equation is equal to zero. Given the short time frame of pressure-decay test (typically 5–10 minutes), it is not clear whether this condition could be met. If the membrane surface in contact with air is not completely dry, the resolution calculation using Equation 5-1 would not be applicable and the contact angle measured using dry membrane media (e.g., sessile drop and the Wihelmy method) would not be relevant. In this case, the contact angle measured by the captive bubble method may be the most proper one because the condition used for measurement is close to the actual condition under which the pressure-decay test is performed.
- Water contact angle is a statistical average and is subject to significant local variations. The measurement of contact angle can be affected by many factors (Cheryan 1998). The local heterogeneity of membrane medium can have significant impact on contact angle measurement. Further, the resolution criterion is supposed to represent the worst-case scenario and should be based on a conservative estimate.
- It has been reported that the adsorption of natural organic matter could change a membrane's water contact angle (Childress and Elimelech 1996; Jucker and Clark 1994, Lozier et al. 2008). Therefore, using the water contact angle measured with a virgin membrane in Equation 5-1 may not represent the real-world condition and can potentially obtain erroneous results.
- The measurement of the water contact angle is a rather complicated task and the results can be variable also, depending on the person performing the test. It is unrealistic to expect that the test is performed routinely by a plant operator. In fact, water contact angle is not even routinely measured by membrane manufacturers.
- Similarly, the value of κ is uncertain because there is no practical means to predict what the shape of a membrane breach would be.

The USEPA's MFGM states that unless it is "scientifically defensible," the values for κ and θ should be considered to be 1 and zero, respectively, on account of being conservative. Given the uncertainties previously discussed, the use of conservative values of θ and κ is to err on the safe side. A sensitivity analysis was conducted to evaluate the impact of incorporating uncertainties in the calculation of resolution of the direct integrity testing (Minnery et al. 2009). Using Monte Carlo and Probability Bounds Analysis, researchers found that resolution may exceed 3 μm and may be as large as 8 μm. With a variable contact angle model, all lower bound possibilities met the resolution criterion for five commercial membrane systems, whereas the upper bound is over 80 percent noncompliant for three of five commercial systems.

DIFFUSION OF AIR IN WATER ACROSS AN INTEGRAL MEMBRANE MODULE

When membrane is integral, the only route for air to pass through a membrane is through diffusion. Quantifying the air diffusion through a membrane is important to determine the baseline value for any integrity testing method based on the flow of fluid, as illustrated in Figure 5-1. This section describes the theoretical calculation based on a simple model and the compares the results of calculation with actually measured value.

Consider a typical hollow-fiber membrane undergoing a pressure-decay test as illustrated in Figure 5-3 in which the module is pressurized from the shell side. Fiber lumens are filled with water and the filtrate side of the module is open to the atmosphere. There are two water–air interfaces: one is on the shell side with an air pressure (p_1); the other is on the top of the fiber lumen open to the atmosphere pressure (p_2). Because the shell side of the module is pressurized above atmospheric pressure, the concentration of the dissolved air on this air–water interface is higher than that on the filtrate-side air–water interface. Therefore, there is a concentration gradient of dissolved air across the module with the direction from the shell side to filtrate side. This gradient is the driving force for air diffusion across the module. Air diffusion across the module is described by Fick's First Law:

$$J = D\,\frac{dC}{dx} \qquad \text{(Eq 5-2)}$$

Where

J = air flux, mol/min-M^2

D = diffusivity of air in water, M^2/min

$\frac{dC}{dx}$ = dissolved air concentration gradient across the module, mol/M^4

If air and water are in equilibrium and water vapor pressures are assumed to be negligible, the concentrations of the dissolved air on two air–water interfaces, C_1 and C_2, are expressed as the following according to the Henry's Law:

$$C_1 = p_1\,\mathrm{H} \qquad \text{(Eq 5-3)}$$

$$C_2 = p_2\,\mathrm{H} \qquad \text{(Eq 5-4)}$$

Where

H = Henry's constant (mol/L-atm)

C_1 and C_2 = dissolved air concentrations on shell-side and filtrate-side interfaces (mol/L), respectively

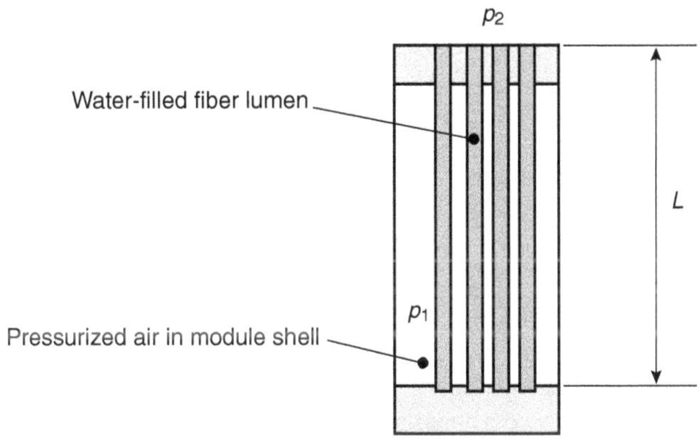

Figure 5-3. Schematic of air diffusion through a hollow-fiber membrane module (p_1 = testing air pressure; p_2 = atmosphere pressure; L = length of diffusion distance)

At the steady-state, at any given time the air coming out from the air–water interface on the filtrate side is equal to the air diffused through the water phase inside the fiber lumen, which, in turn, is equal to the air dissolved into the water on the water–air interface on the shell side of the module. If the concentration distribution along the fiber lumen is known, the air flow via diffusion can be calculated by integrating Equation 5-2:

$$q_{A0} = \varepsilon A J = \varepsilon A \frac{D}{L} \int_0^L C(x)dx \qquad \text{(Eq 5-5)}$$

Where

q_{A0} = airflow via diffusion
A = surface area of module, M^2
ε = the membrane surface porosity
L = length of diffusion distance
$C(x)$ = concentration distribution of dissolved air along the lumen length

Equation 5-5 can be approximated by using a simplified form if it is assumed that concentration distribution is linear along the lumen length and the diffusion within the membrane pores is negligible when comparing it to the diffusion through the fiber lumen if one side of hollow-fiber membrane is filled with water. The second assumption is reasonable considering the length of fiber is more than three orders of magnitude larger than the thickness of the membrane, even though the path of diffusion inside the membrane may be very tortuous. Then, the average gradient and diffusion length can be used to substitute the terms in Equation 5-5.

$$q_{A0} = \varepsilon AD \frac{C_1 - C_2}{(L/2)} \qquad \text{(Eq 5-6)}$$

Substitute Equations 5-3 and 5-4 into 5-6:

$$q_{A0} = 2\varepsilon ADH\frac{p_1 - p_2}{L} \qquad \text{(Eq 5-7)}$$

In Equation 5-7, the diffusivity of air in water, D, can be calculated by the following Wilke-Chang Equation:

$$D = 1.173 \times 10^{-16}\,(\theta M_{H_2O})^{1/2}\frac{T}{\mu_W V_A{}^{0.6}} \qquad \text{(Eq 5-8)}$$

Where

θ = association coefficient of the solvent, φ = 2.6 for water
M_{H_2O} = the molecular water of water (18), g/mol
μ_W = viscosity of water, pa-second
T = absolute temperature (°K)
V_A = solute molar volume at the boiling point, V_A = 0.00299 M^3/Kmol for air

The Henry's constant, H, is also a function of the temperature:

$$\mathrm{H}(T) = \mathrm{H}_{298}e^{\left[-k\left(\frac{1}{T}-\frac{1}{298}\right)\right]} \qquad \text{(Eq 5-9)}$$

Where

H_{298} = Henry's constant at 298°K; H_{298} = 1.4×10^{-3} for air
k = temperature coefficient; k = 1,300 for air.

Laboratory tests using the setup illustrated in Figure 5-4 were conducted to measure the air flow via diffusion of an integral hollow-fiber membrane module (Liu 1998). The results are plotted in Figure 5-5 and compared with theoretical calculation by Equation 5-7. The results in Figure 5-5 show that air diffusion rate calculated by Equation 5-7 matched well to the measured values.

Both theoretical calculation and experimentation indicate that the diffusion of air across a hollow-fiber membrane module is relatively small. As a result, the pressure decay caused by the diffusion would not be observable. In reality, low pressure decays are typically observed. However, this is most likely the results air leakage from seals and piping system, rather than as a result of air diffusion through the membrane. Another possible source of pressure decay is the displacement of water by air on the side of membranes where air is pressurized. If the amount of water is not displaced at the beginning of the pressure-decay test, pressure decay caused by water displacement would be counted.

It should be noted that Equations 5-6 and 5-7 were developed based on the assumption that one side of the membrane is filled with water as illustrated in Figure 5-3. This is the most likely case for membrane integrity testing in practice. However, if both sides

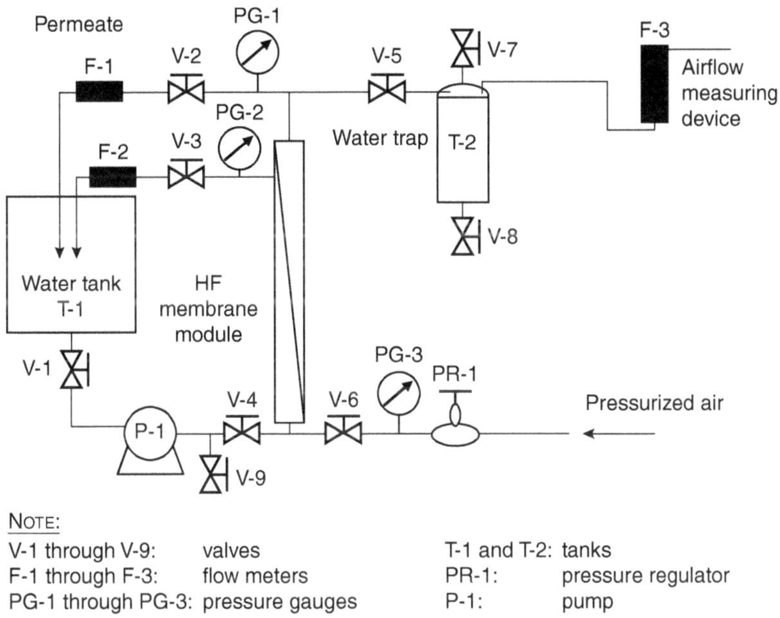

Figure 5-4. Schematic of experimental setup to measure air diffusion flow through a hollow-fiber membrane module

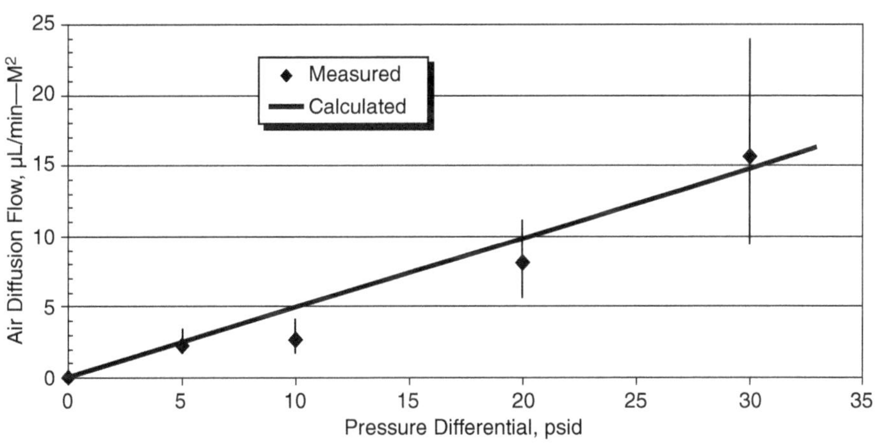

Figure 5-5. Comparison of measured and calculated air diffusion flow using Equation 5-7 [*D* (fiber inner diameter) = 0.0006 m; *L* (fiber length) = 1.5 m; *t* (temperature) = 21°C]

of the membrane are filled with air, the length of the diffusion path is reduced to the length of the diffusion path within the membrane walls. In this case, the diffusive air flow (q_{A0}) would be considerably higher than that predicted by Equation 5-7 due to a much shorter diffusion path. In pressure-decay test, the pressure decay rate caused by air passage from diffusion is extracted from measured pressure decay rate to calculate log reduction value (LRV) of a membrane system.

Table 5-1. Comparison of hydraulic modeling and empirical correlation for LRV determination

Hydraulic Modeling	Empirical Correlation
• Simplified version of real world	• Actual measurements not relying on assumptions
• Preferably erring on the side of conservatism (predicting lower LRV)	• Can be systematic and/or site-specific
• Good for quantitative analysis	• The range of applicability is limited by the conditions under which measurements are made
• Validation of the model is required	• Results may be influenced by noncontrol variables

CALCULATION OF LRV

There two general approaches to calculate LRV: hydraulic modeling and empirical correlation. The former depends on the equations describing hydraulics of both water and air passage through a given membrane breach, the latter is based on the actual measurements of water and air passages and correlation of two for the given membrane breach. The features of those two different approaches are presented in Table 5-1. Naturally, two approaches can also be combined to yield the results of LRV calculation. There are several published methods for LRV calculation (Liu and Schafer 2002; ASTM 2003; USEPA 2005). The following sections describe those methods and a comparison will be made at the end of the chapter.

ASTM D-6908-03 Method

ASTM D-6908-03: "Standard Practice for Integrity Testing for Water Filtration Membrane Systems" (referred to as "ASTM practice" hereafter), published in 2003, includes the pressure-decay test (PDT) and vacuum-decay test (VDT) as Practice A. This practice may be used on membranes of all classes and configurations. However, in practice, PDT is most commonly used for UF and MF, while VDT is used for testing NF and RO. The major difference between PDT and VDT is that the pressure differentials across the membranes are positive for PDT and negative (vacuum) for VDT, respectively. Therefore, for PDT, air can be on one side or both sides of the membrane, while air must be on both sides of membrane for VDT, as depicted in Figure 5-6. As the topic is integrity testing of low-pressure membranes, the focus of the discussion will be on PDT.

The ASTM practice specifies that the selection of pressure is based on Cantor's Equation and does not include shape factor as in the USEPA's MFGM. The ASTM practice also provides two methods for calculation of LRV: (1) calculation of LRV

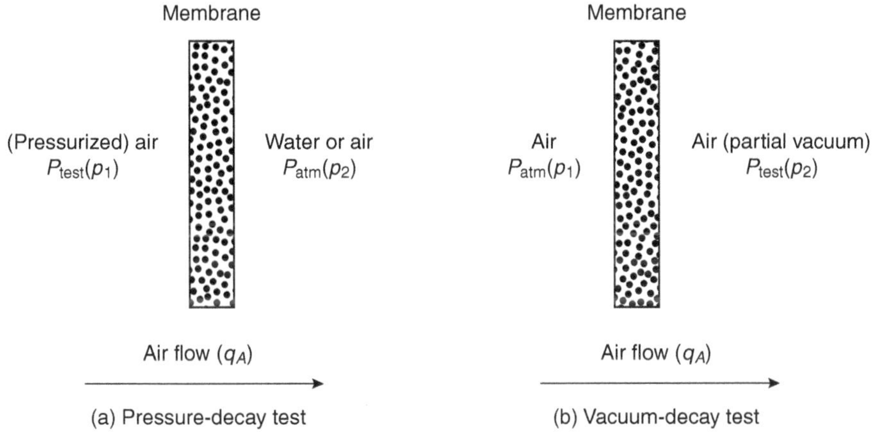

Figure 5-6. Comparison of test schematic for pressure-decay test and vacuum-decay test

using Hagen-Poiseuille Equation, and (2) calculation of LRV using empirical correlation between PDR and bypass flow (q_W) through fiber-cut test.

LRV Calculation Using Hagen-Poiseuille (H-P) Equation

This method assumes both air and water flow through a cylinder-shaped membrane breach in laminar regime. The major equations for this method are presented below:

$$\mathrm{LRV} = \log\left(\frac{QP_{atm}}{\mathrm{VCF} \times \mathrm{PDR} \times V_H} f_1 f_2\right) \qquad \text{(Eq 5-10)}$$

Where

Q = system filtrate flow
VCF = volume concentration factor
PDR = pressure-decay rate
V_H = system hold-up volume (the volume occupied by pressurized air during pressure-hold test)
P_{atm} = atmosphere
f_1 = viscosity correction factor
f_2 = pressure correction factor

f_1 and f_2 are defined as the following:

$$f_1 = \frac{\mu_W}{\mu_A} \qquad \text{(Eq 5-11)}$$

Where

μ_W = viscosity of water
μ_A = viscosity of air

$$f_2 = \frac{p_1^2 - p_2^2}{2P_{atm}\Delta P} \quad \text{(Eq 5-12)}$$

Where

ΔP = trans-membrane pressure during the filtration phase

MFGM Method

The centerpiece of MFGM method is to determine air–liquid conversion ratio (ALCR) as defined as following:

$$\text{ALCR} = \frac{q_A}{q_W} \quad \text{(Eq 5-13)}$$

This approach holds the promise that air flow and water bypass flow through the defect can be related via ALCR. The key task is then to convert the result from flow-based integrity testing (e.g., pressure-decay test) to the bypass flow from the membrane breach via ALCR. This section presents the key elements of the LRV calculation in MFGM. For detailed information, the reader should refer to Appendices C in the MFGM (USEPA 2005).

Calculating ALCR Via Mathematical Modeling

To accommodate the various membrane configurations and the types of membrane breaches, MFGM proposes a generalized approach for calculating ALCR, as described in Table 5-2. The MFGM approach divides the calculation into two membrane configurations: tubular and flat sheet. The former includes hollow-fiber and tubular membrane filters, whereas the latter includes spiral wound and pleat filters. Each membrane configuration is then subdivided into two flow regimes: turbulent and laminar. For turbulent regime, the equations for Darcy pipe flow and for flow through an orifice are used for tubular and flat sheet configurations, respectively. For laminar regime, Hagen-Poiseuille Equation is used regardless of membrane configuration.

Table 5-2. MFGM approaches for calculating ALCR (USEPA 2005)

Membrane Configuration	Flow Regime Through Breach	Model for Calculating ALCR
Hollow fiber and tubular	Turbulent	Darcy pipe flow
	Laminar	Hagen-Poiseuille Equation
Flat sheet (including spiral wound and pleat filters)	Turbulent	Flow through an orifice
	Laminar	Hagen-Poiseuille Equation

MFGM = *Membrane Filtration Guidance Manual*
ALCR = air–liquid conversion ratio

Darcy pipe flow. This equation describes the membrane breach as a completely broken hollow-fiber or tubular membrane or as large-diameter hollow-fiber or tubular membrane with short flow path and high pressure differential. The method assumes that both air flow and water bypass flow are in a turbulent regime. As such, the air flow and bypass flow are described by Darcy pipe flow equations for compressible and incompressible fluid, respectively:

$$q_A = 11.3 \times Y \times D^2 \left(\frac{(p_1 - p_2)p_1}{(460 + t)xK_A} \right) \quad \text{(Eq 5-14)}$$

$$q_W = 0.525 \times D^2 \left(\frac{\Delta P}{\rho_W \times K_W} \right)^{0.5} \quad \text{(Eq 5-15)}$$

Where

Y = net expansion factor for compressible fluid (air), dimensionless
t = water temperature, °F
K_A = resistance coefficient of air, dimensionless
p_1, p_2 = air pressure at the inlet and the out let of the membrane breach, psia
K_W = resistance coefficient of water, dimensionless
ρ_W = density of water, lb/ft^3
ΔP = pressure differential or trans-membrane pressure, psi
q_A, q_W = air flow and water bypass flow through the membrane breach, ft^3/s, respectively; air flow refers to that under the standard condition (60°F and 14.7 psia)

It should be noted that Equations 5-14 and 5-15 have consistently resulted from unit conversion. Therefore, the units of each term have to be as indicated.

By the definition of ALCR the result is:

$$\text{ALCR} = \frac{q_A}{q_W} - 21.52 \times Y \times \left(\frac{(p_1 - p_2)p_1}{(460 + t)\times \Delta P} \frac{K_W}{K_A} \rho_W \right)^{0.5} \quad \text{(Eq 5-16)}$$

Assuming $K_A = K_W$ and apply a value of 62.4 lb/ft^3 for ρ_W, Equation 5-16 becomes:

$$\text{ALCR} = 170 \times Y \times \left(\frac{(p_1 - p_2)p_1}{(460 + t) \times \Delta P} \right)^{0.5} \quad \text{(Eq 5-17)}$$

The net expansion factor (Y) for air takes the change in air volumetric flow into account as the density of air changes with the pressure. The value of Y is a function of both the pressure differential of air flow and resistance coefficient (K_A). Some references for hydraulics provide charts. Resistance coefficient, K_A, is defined as follows:

$$K_A = f \frac{L}{D} \quad \text{(Eq 5-18)}$$

As indicated in Equation 5-18, the resistance coefficient is a function of both the geometry of the membrane breach and friction factor. In consideration of the worst-case scenario of the maximum bypass flow, the MFGM suggests that the shortest flow path should be used in determining K_A's value. As for friction factor, *f*, its value can be determined through iterations by using the Colebrook Equation (Giles et al. 1995):

$$\frac{1}{\sqrt{f}} = 2 \log\left(\frac{\varepsilon}{3.7D} + \frac{2.51}{\text{Re}\sqrt{f}}\right) \qquad \text{(Eq 5-19)}$$

Where

ε/D = relative roughness of hollow-fiber lumen

Flow through an orifice. This method is used to calculate ALCR for breaches on flat sheet membranes, such as in spiral wound or pleat cartridge filter elements. Again, the method assumes that both air flow and water bypass flow are in a turbulent regime. The equations for air flow and water bypass flow are the following:

$$q_A = 11.3 \times Y \times D^2 \times C\left(\frac{(p_1 - p_2)p_1}{(460 + t)}\right)^{0.5} \qquad \text{(Eq 5-20)}$$

$$q_W = 0.525 \times D^2 \times C\left(\frac{\Delta P}{\rho_W}\right)^{0.5} \qquad \text{(Eq 5-21)}$$

where C = coefficient of orifice discharge, dimensionless; and all other elements are the same as in Equations 5-14 and 5-15.

Again, taking the value of ρ_W as 62.4 lb/ft^3, ALCR can be calculated in:

$$\text{ALCR} = 170 \times Y\left(\frac{(p_1 - p_2)p_1}{(460 + t)\Delta P}\right)^{0.5} \qquad \text{(Eq 5-22)}$$

Equation 5-22 is identical with Equation 5-17. However, the net expansion factor in Equation 5-22 is calculated differently:

$$Y = 1 - \left[0.293\left(1 - \frac{p_2}{p_1}\right)\right] \qquad \text{(Eq 5-23)}$$

Laminar flow. This equation is used for the membrane breach of either a completely broken fiber or a hole on a flat sheet membrane. The equation to calculate air flow in laminar regime in the MFGM is the Hagen-Poiseuille (H-P) Equation:

$$q_A = \left(\frac{49.5\pi D^4\, \Delta P^*\, g}{L \times \mu_A \times (460 + t)}\,\frac{p_2}{P_{atm}}\right) \qquad \text{(Eq 5-24)}$$

Where

q_A = air flow through the membrane breach under the standard conditions (60°F and 14.7 psi), ft^3/s

D, L = diameter of hollow-fiber lumen and the length of flow path, respectively, in.

p_2, P_{atm} = pressure at lumen outlet and atmosphere, respectively, psia

μ_A = viscosity of air, lbf/ft-s

t = temperature, °F

g = gravity acceleration, g = 32.2 lbm-ft/lbf-s^2

ΔP^* = effective testing pressure, as expressed as the following:

$$\Delta P^* = (p_1 - p_2)\,\frac{p_1 + p_2}{2p_2}\,\frac{p_2}{P_{atm}} \qquad \text{(Eq 5-25)}$$

Water bypass flow through the membrane breach:

$$q_W = \frac{0.094\pi D^4\,\Delta P}{L \times \mu_W} \qquad \text{(Eq 5-26)}$$

By combining Equations 5-28 and 5-30, the result is:

$$\text{ALCR} = \frac{527\Delta P^* \times \mu_W}{\Delta P \times (460 + t) \times \mu_A} \qquad \text{(Eq 5-27)}$$

Once ALCR is calculated via the proper equation, the log reduction value can be calculated as following:

$$\text{LRV} = \log\left(\frac{Q \times P_{atm} \times \text{ALCR}}{\text{VCF} \times \text{PDR} \times V_H}\right) \qquad \text{(Eq 5-28)}$$

If we substitute Equations 5-25 and 5-27 into Equation 5-28,

$$\begin{aligned}\text{LRV} &= \log\left[\frac{527}{460 + T}\left(\frac{Q \times P_{atm}}{\text{VCF} \times \text{PDR} \times V_H}\right)\left(\frac{(p_1 - p_2)(p_1 + p_2)}{2p_2}\,\frac{p_2}{P_{atm}}\right)\left(\frac{\mu_W}{\mu_A}\right)\right] \\ &= \log\left[\frac{527}{460 + T}\left(\frac{Q \times P_{atm}}{\text{VCF} \times \text{PDR} \times V_H}\right)\left(\frac{(p_1 - p_2)^2}{2p_{atm}}\right)\left(\frac{\mu_W}{\mu_A}\right)\right]\end{aligned} \qquad \text{(Eq 5-29)}$$

When compared to the equation of ASTM method (Equation 5-10), it should be noted that they are really identical equations if unit conversion factors are taken into the account. Considering both equations are based on the Hagen-Poiseuille Equation, the fact that both have the same form should be expected.

An Alternative Method to Calculate LRV—Pall Equation

An alternative method to the ASTM and MFGM methods for calculating LRV discussed previously is presented here (Liu and Schaefer 2002). The method includes several steps:

1. To develop an equation for air flow through the membrane defect (defined as a completely broken hollow fiber)
2. To relate air flow to the pressure-decay rate
3. To develop an equation for water bypass flow through the membrane defect
4. To relate water bypass flow to LRV
5. To relate pressure decay to LRV

A key feature is that the equation describing air passage through the membrane defect (completely broken fiber) is derived from the energy conservation and mass conservation of compressible fluid, and does not require the assumption of flow regime.

Air Passage Through a Breach Hollow Fiber

A hollow-fiber membrane module that has a breached fiber is used as an example. Figure 5-3 offers a similar example, but L is defined as the length of fiber lumen from the point of breach to the lumen's outlet. The equation can be written for energy conservation (i.e., the Bernolli Equation) to the hollow-fiber segment allowing the following assumptions:

- Air forms a continuous, steady one-phase flow in a hollow fiber and resembles that in an impermeable pipe. The diffusion of air from and/or into the fiber lumen through fiber walls is negligible.
- Air behaves like ideal gas.
- Isothermal condition is assumed in the fiber lumen, i.e., temperature in fiber lumen does not change along the flow path or over time.
- Minor hydraulic losses of air flow are negligible compared to losses from the fiber lumen.

Noting that air is a compressible fluid, the volumetric velocity terms in the equation need to be replaced with the mass velocity. Then, the result is:

$$\frac{M}{2RT}(p_2^2 - p_1^2) + U_M^2 \ln\frac{V_{S2}}{V_{S1}} + \frac{fU_M^2L}{2D} = 0 \qquad \text{(Eq 5-30)}$$

Where

M = molecular weight of air, kg/mol

R = gas constant, R = 8.314 Pa-M^3/gmol-°K

U_M = mass velocity, kg/s–M^2

V_S = specific volume of air, M^3/kg

f = friction coefficient, dimensionless

D = inner diameter of the hollow-fiber lumen

p = air pressure

The subscripts 1 and 2 represent the parameters at the entrance and at the exit of the fiber lumen, respectively.

Based on the equation of energy conservation (Equation 5-30), the velocity of airflow at the inlet of the hollow fiber can be derived as follows (see appendix 5-A for mathematical derivation):

$$v_1 = \left[\frac{1}{\rho_1 \left(\frac{fL}{D} + 2\ln \frac{p_1}{p_2} \right)} \frac{(p_1^2 - p_2^2)}{p_1} \right]^{1/2} \qquad \text{(Eq 5-31)}$$

Where

v_1 = velocity of airflow at the inlet of the hollow fiber

p_1, p_2 = (absolute) pressure at the inlet and the outlet of the hollow fiber, respectively

ρ_1 = air density at the inlet of the hollow fiber

f = friction factor

Because air is treated as ideal gas, the velocity at the outlet of a hollow fiber can be calculated as following:

$$v_2 = \frac{p_1}{p_2} v_1 \qquad \text{(Eq 5-32)}$$

Airflow at the outlet can be calculated as following:

$$q_A = Av_2 \qquad \text{(Eq 5-33)}$$

where $A = \pi D^2/4$, the area of the cross-section of the hollow-fiber lumen.

Substitute Equations 5-31 and 5-32 into Equation 5-33:

$$q_A = A\frac{p_1}{p_2}\left[\frac{1}{\rho_1 \left(\frac{fL}{D} + 2\ln \frac{p_1}{p_2} \right)} \frac{(p_1^2 - p_2^2)}{p_1} \right]^{1/2} \qquad \text{(Eq 5-34)}$$

In Equation 5-34, the value of friction factor is a function for both Reynolds number and toughness of fiber lumen. An example of friction coefficient as a function Reynolds number is illustrated in Figure 5-7 for a fixed relative toughness of 1.1×10^{-5}.

Relating the Air Flow to Pressure Decay

If the volume occupied by the air during a pressure-decay test (also as known as hold-up volume) is known, the pressure-decay rate (PDR) can be calculated as following:

$$\text{PDR} = q_A \frac{\rho_2 RT}{MV_H} \qquad \text{(Eq 5-35)}$$

Where

V_H = volume occupied by air during the test (hold-up volume)

The derivation of Equation 5-35 is shown in appendix 5-B.

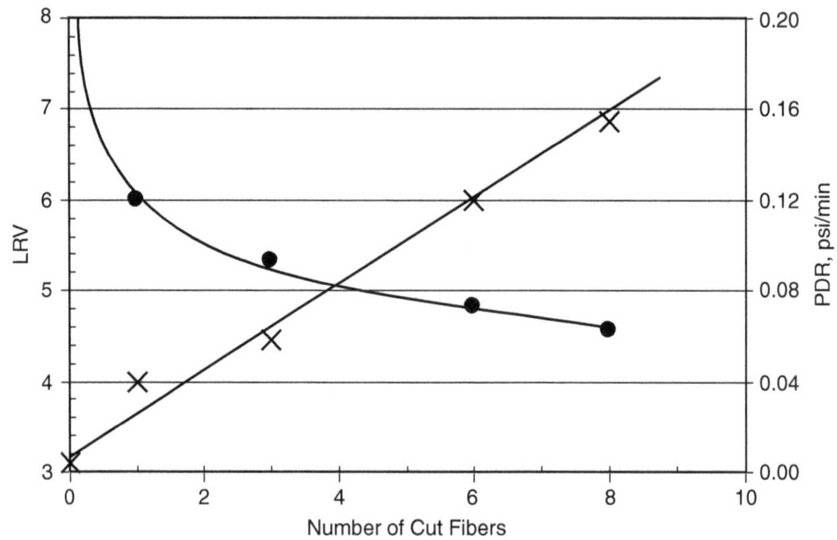

LRV = log reduction value; PDR = pressure-decay rate

Figure 5-7. LRV and PDR as a function of the number of cut fibers (plotted based on data of Sethi et al. 2004)

Substitute Equation 5-34 into Equation 5-35 and rearrange to solve for A (note that $p/\rho = RT/M$)

$$A = \left(\frac{M}{RT}\right)^{0.5}\left(\frac{f\dfrac{L}{D} + 2\ln\dfrac{p_1}{p_2}}{p_1^2 - p_2^2}\right)^{0.5} \qquad \text{(Eq 5-36)}$$

Water Bypass Flow Through a Breached Hollow-Fiber Membrane

The bypass flow through a completely breached fiber is discussed extensively in chapter 3. In this chapter, only the highlights are provided. Similar to the section for discussing air flow through a breached hollow fiber, the bypass flow can be derived through the principles of hydraulics such as mass and energy conservation, namely, the Bernoulli Equation. If the minor losses at the entrance and the exit of the hollow fiber are negligible compared to major loss in the fiber lumen and the fiber lumen is considered to be hydraulically smooth, the equations in Table 3-3 can be employed for hydraulic calculations of bypass flow in an outside-in mode for two ranges of values of Reynolds number (Re ≤ 2,300 and 4,000 < Re < 100,000). When flow regime is laminar (Re ≤ 2,300), the relationship between water bypass flow and pressure differential is linear.

Outside the laminar regime, the relationship between bypass flow and pressure differential is no longer linear due to increased hydraulic resistance to the flow. At a high Reynolds number, the bypass flow as a function of pressure differential is more likely to follow the trend calculated with equations for nonlaminar flow regime, as illustrated in Figure 3-3. Therefore, bypass flow calculated from Equation 3-9 provides a conservative estimate:

$$q_W = \frac{\pi \Delta P D^4}{128 \mu_W L} = \frac{\pi D^2}{4} \frac{\Delta P D^2}{32 \mu_W L} = A \frac{\Delta P D^2}{32 \mu_W L} \qquad \text{(Eq 5-37)}$$

Relating LRV to Water Bypass Flow

LRV can be related to water bypass flow by substituting Equation 3-9 into Equation 3-3:

$$\text{LRV} = \log \frac{Q}{q_W \times \text{VCF}} = \log \frac{32 Q \mu_W L}{A \Delta P D^2 \times \text{VCF}} \qquad \text{(Eq 5-38)}$$

Relating Pressure Decay to LRV

Substitute Equation 5-36 into Equation 5-38 and rearrange:

$$\text{LRV} = 1.505 + \log \frac{Q_W \mu_W}{V_H \Delta P \times \text{VCF}} + \log \left(\frac{RT}{M}\right)^{0.5} \frac{(p_1^2 - p_2^2)^{0.5}}{D(\text{PDR})} + \log \left\{ \frac{\dfrac{L}{D}}{\left[\dfrac{fL}{D} + 2\ln\left(\dfrac{p_1}{p_2}\right)\right]^{0.5}} \right\} \qquad \text{(Eq 5-39)}$$

Where

Q_W = system filtrate flow
μ_W = water viscosity
V_H = system hold-up volume (the volume occupied by pressurized air during pressure-hold test)
ΔP = pressure differential across the system, approximated by transmembrane pressure
VCF = volume concentration factor
R = universal gas constant
T = temperature
M = average molecular weight of the air
p_1, p_2 = pressures at upstream and downstream of the membrane
D = inner diameter of the hollow-fiber lumen
PDR = pressure-decay rate
L = length of flow path in the hollow-fiber lumen
f = friction factor of air flow in the hollow-fiber lumen

The last three terms in the equation are arranged to be dimensionless. Therefore, any set of units can be applied for each of three terms as far as they are uniform. Equation 5-39 assumes the following:

- Water flow in the hollow-fiber lumen is laminar.
- The flows (both water and air) in the lumen of the broken hollow fiber can be approximated with the flow in an impermeable pipe, i.e., the flows though

membrane walls of the hollow fiber are far smaller than the flow through the broken fiber.

- Air behaves like an ideal gas and isothermal conditions are maintained in the hollow-fiber lumen.
- Minor local hydraulic losses such as occurred at the entrance and exit of the breached hollow fiber are neglected.

While being conservative, those conditions are considered to be reasonably representative of the real world. It should be noted that the calculation can be expanded into the transition and turbulent regimes. Under such a situation, the relationship between the friction coefficient and Reynolds number is no longer explicit and the value of friction coefficient is also dependent on the relative roughness. One would need to do iteration to obtain the proper value for friction coefficient.

The relationship between LRV and PDR is rather useful to set the control limit for integrity testing. This enables the plant operator to make a decision on when the plant needs to be shut down for repairing or replacing the membranes.

SENSITIVITY ANALYSIS

Equations for LRV calculation and major assumptions for each are summarized in Table 5-3. A major advantage to having a theoretical model is the ability to conduct sensitivity analysis of various parameters and to examine how they would affect LRV. There are five parameters selected for the evaluation for their impact on LRV: (1) temperature (t), (2) friction factor (f), (3) the geometry of the membrane breach as defined by the ratio of length to fiber diameter (L/D), (4) the testing pressure (p_1), and (5) the hold-up volume (V_H). Those parameters are chosen to represent environmental conditions (1); system characteristics—(2), (3), and (5); and integrity testing conditions (4), respectively. After setting the baseline value, a change of the selected parameter as a percentage of the baseline value is chosen for each parameter. Baseline value and the range of change in each parameter are summarized in Table 5-4. The range of the change generally covers the range of the parameters that may be encountered in practice. The changes in LRV are computed using those equations for each parameter with changes below and above the baseline value.

Because the equation for the ASTM method (Equation 5-10) and the equation for laminar flow in the MFGM (Equation 5-29) are identical, sensitivity analysis is performed only using Equation 5-29. In addition, friction factor and the geometry of the membrane breach are not performed in the sensitivity analysis for the equations for the turbulent flow of orifice and laminar flow in the MFGM as they are not included in those equations.

The results, as expressed as the sensitivity score (the average change in LRV value per percent change to the baseline value of each parameter) are summarized

Table 5-3. Summary of Equations for Calculating LRV

Equation	Expression	Assumptions / Conditions*
ASTM	$\mathrm{LRV} = \log\left(\dfrac{Q}{\mathrm{VCF} \times \mathrm{PDR} \times V_H} \dfrac{\mu_W}{\mu_A} \dfrac{p_1^2 - p_2^2}{2\Delta P}\right)$	• Membrane breach is broken hollow fiber(s) • Both air and water bypass flows follow Hagen-Poiseuille Equation • Average pressure during the integrity testing is used for p_1
MFGM Darcy	$\mathrm{LRV} = \log\left[\dfrac{Q \times P_{atm}}{\mathrm{VCF} \times \mathrm{PDR} \times V_H} 170 \times Y \times \left(\dfrac{(p_1 - p_2)p_1}{(460 + t) \times \Delta P}\right)^{0.5}\right]$ †	• Membrane breach is a round hole on a flat sheet membrane or a completely broken hollow fiber • Both air and water bypass flows are in turbulent regime • Both air and water bypass flows follow the equation for Darcy pipe flow • $K_A = K_W$ • Change in p_1 is negligible and does not affect pressure-decay rate
MFGM H-P	$\mathrm{LRV} = \log\left(\dfrac{Q}{\mathrm{VCF} \times \mathrm{PDR} \times V_H} \dfrac{527\ \mu_W}{\mu_A(460 + t)} \dfrac{(p_1^2 - p_2^2)}{2\Delta P}\right)$	• Membrane breach is a round hole on a flat sheet of membrane or a completely broken hollow fiber • All others the same as ASTM equation
Pall	$\mathrm{LRV} = 1.505 + \log\dfrac{Q\,\mu_W}{V_H\,\Delta P \times \mathrm{VCF}} + \log\left(\dfrac{RT}{M}\right)^{0.5} \dfrac{(p_1^2 - p_2^2)^{0.5}}{D(\mathrm{PDR})} + \log\left\{\dfrac{\dfrac{L}{D}}{\left[\dfrac{fL}{D} + 2\ln\left(\dfrac{p_1}{p_2}\right)\right]^{0.5}}\right\}$	• Membrane breach is broken hollow fiber(s) • Water bypass flow is laminar • Air behaves like ideal gas (compressible) • Change in p_1 is negligible and does not affect pressure-decay rate

*Common assumptions/conditions for all equations:

- Isothermal flows
- Permeation through membranes is negligible compared with bypass flows
- Minor losses of water bypass flow are negligible

†For completely broken hollow fiber, Y is a function of p_1/p_2 and resistance coefficient, K. The value of Y is solved graphically.

For a hole on a flat sheet, Y is solved by the following: $Y = 1 - 0.293\left(1 - \dfrac{p_2}{p_1}\right)$

Table 5-4. Typical ranges of various parameters related to sensitivity analysis

Parameter	Unit	Range
Temperature (t)	°C	0–20
Friction factor (f)	—	0.02–0.08
Geometry of the membrane breach (L/D)	—	100–300
Testing pressure (p_1)	psi	10–30
Hold-up volume (V_H)	gal	220–440

in Table 5-4. Based on the sensitivity score (the average change in LRV over the percent change in a parameter), the parameters are ranked for their impacts on LRV calculation in the following descending order: hold-up volume (V_H), testing pressure (p_1), geometry of the breach (L/D), friction factor (f), and temperature. An exception is the equation of laminar flow in the MFGM in which testing pressure has the greatest impact among all parameters. Depending on the specific parameter and equation used for the sensitivity analysis, the changes in LRV range from less than 0.01-logs to 0.66-logs over the change in parameters (33–100 percent). This is not unexpected because LRV is based on logarithm value, which "damps" the impact of those parameters. That is, a factor of 2 in arithmetic value only yields a value of 0.3 in logarithm scale. The parameters selected for the sensitivity analysis can have either a positive (i.e., increasing the value) or a negative (i.e., decreasing the value) impact on LRV, as indicated in Table 5-5. Among them, geometry of the membrane breach (L/D) and testing pressure (p_1) have a positive impact and the other three variables have a negative impact.

The sensitivity of different equations for the same parameter can be different. For example, a change in temperature of 10°C yields only a change of less than 0.01-logs in resulting LRV for the equations of hollow fiber and orifice under turbulent conditions in the MFGM. However, the same change yields a change of 0.26-logs in LRV for the equation for laminar flow in the MFGM. Another example is the impact of friction factor and the geometry of the membrane breach. For the equation for hollow fiber in turbulent flow in the MFGM, friction factor and the geometry of the membrane breach have only a minor impact on LRV (~0.04-logs). However, the equation of the alternative method is much more sensitive to the change in fraction factor and the geometry of the membrane breach, resulting in changes in LRV of 0.20- and 0.28-logs, respectively.

It is helpful to understand the impact of each parameter if we analyze what each parameter means physically. The increase in L/D means either the increase in flow path for fluids from the breach, or smaller hollow-fiber lumen—both would mean more restriction to the flow. As a result, bypass flow is reduced and LRV increases as the value of L/D increases. Because the diameter of hollow-fiber lumen is generally

Table 5-5. Sensitivity analysis for the impacts of various parameters on LRV calculation using various equations

Parameter	Baseline Value	Change (Δx)	Δx%	ΔLRV ($-\Delta x$)	ΔLRV ($+\Delta x$)	Sensitivity Score*	Impact of Increasing Value
MFGM Darcy: Hollow-fiber membrane and turbulent flow							
Temperature (t), °C	10	10	100%	+0.01	−0.01	0.01	Decreasing LRV
Fraction factor (f)	0.05	0.03	60%	−0.04	+0.01	0.04	Increasing LRV
Geometry of breach (L/D), dimensionless	200	100	50%	−0.03	+0.01	0.04	Increasing LRV
Testing pressure (p_1), psi	20	10	50%	−0.19	+0.12	0.31	Increasing LRV
Hold-up volume (V_H), gal	330	110	33%	+0.17	−0.12	0.45	Decreasing LRV
MFGM Darcy: Flat sheet membrane (orifice) and turbulent flow							
Temperature (t), °C	10	10	100%	+0.01	−0.01	0.01	Decreasing LRV
Testing pressure (p_1), psi	20	10	50%	−0.20	+0.13	0.33	Increasing LRV
Hold-up volume (V_H), gal	330	110	33%	+0.17	−0.12	0.45	Decreasing LRV
MFGM Hagen-Poiseuille (H-P): Hollow-fiber or flat sheet membrane and laminar flow							
Temperature (t), °C	10	10	100%	+0.28	−0.24	0.26	Decreasing LRV
Testing pressure (p_1), psi	20	10	50%	−0.40	+0.26	0.66	Increasing LRV
Hold-up Volume (V_H), gal	330	110	33%	+0.17	−0.12	0.45	Decreasing LRV
Pall: Hollow-fiber membrane							
Temperature (t), °C	10	10	100%	+0.12	−0.12	0.12	Decreasing LRV
Friction factor (f), dimensionless	0.05	0.03	60%	+0.16	−0.09	0.20	Decreasing LRV
Geometry of breach (L/D), dimensionless	200	100	50%	−0.18	+0.10	0.28	Increasing LRV
Testing pressure (p_1), psi	20	10	50%	−0.19	+0.12	0.31	Increasing LRV
Hold-up volume (V_H), gal	330	110	33%	+0.17	−0.12	0.45	Decreasing LRV

*Sensitivity score is calculated by the following: $\dfrac{|\Delta\mathrm{LRV}(-\Delta x) - \Delta\mathrm{LRV}(+\Delta x)|}{2\Delta x\%}$

uniform, the change in *L/D* represents the position at which membrane breach occurs. Figure 5-18 (page 97) compares the effect of the flow path, *L*, on LRV by assuming complete breached hollow fiber at the near end, the middle, and the far end from the membrane-potting interface on filtrate side (with a full length of 2 m). It should be noted that *L/D* ratio under such condition is far greater than the change shown in Table 5-1. Therefore, the effect on LRV is more profound. As indicated in Figure 5-18, bypass flow with the shortest flow path when breach occurs near the filtrate end potting can be one to two orders of magnitude higher than when the breach occurs at the middle and the far end of the filtrate end potting, respectively, where the flow path is increasingly longer. Nevertheless, the effect of position of the membrane breach can be considerable.

Increasing testing pressure for a pressure-decay test would increase more air flowing through the same membrane breach. Even though it does not reduce the bypass flow through the breach, it increases the value of PDR. Conversely, it means that the same PDR value from increased testing pressure implies a smaller membrane breach, thus, the higher LRV. The impact of hold-up volume has the opposite effect on LRV as testing pressure. For the same membrane breach and the testing pressure, the air flow through the breach would be the same regardless of the hold-up volume of a membrane system.

However, a system with a larger hold-up volume holds more air, so it takes longer to reduce the air pressure. This would mean a lower PDR. Conversely, if the value of PDR is the same for two membrane systems with different hold-up volumes, the one with a larger hold-up volume would have to have a larger membrane breach (therefore a low LRV).

The impact of temperature is less explicit as it appears in Equation 5-39. Increasing temperature reduces the viscosity of the fluid, which in turn increases bypass flow by reducing the hydraulic resistance (therefore, decreasing LRV). Compared to water, the viscosity of air has less effect in the range of temperature used in the sensitivity analysis. The more profound impact of temperature on air flow is the change in air density. Increasing temperature increases air flow by making air expand. This has a negative effect on LRV. Increasing temperature also increases the diffusivity of air in water, as described in Equation 5-8. This may change the baseline value for pressure-decay test. The overall effect of temperature is negative for LRV.

Similar to the effect of temperature, the friction coefficient affects both water bypass flow and air flow. Because it is assumed that the bypass flow is in laminar regime, the friction coefficient for bypass flow is an explicit function of the Reynolds number. As a result, the term only appears for air flow. Increasing the value of friction coefficient reduces air flow but not bypass flow according to Equation 5-39. Therefore, LRV decreases as friction coefficient increases.

It should be noted that friction factor can be a function of Reynolds number alone, or of both Reynolds number and the roughness of the hollow-fiber lumen, or of the roughness of hollow-fiber lumen alone, depending on the flow regime. To get the exact value of friction coefficient, one needs to determine if the flow is in transition regime. The range of the value used for sensitivity analysis is rather wide for practical purposes except perhaps in the case of hydraulically very smooth lumen walls of hollow-fiber membranes. The result of sensitivity analysis indicates that a 60 percent change in the value of friction factor from the baseline value only yields a moderate change in LRV about two tenths of a log below and above the baseline LRV, respectively.

The magnitude of the impact of friction factor on LRV also depends on the *L/D* ratio. The impact of friction factor can also be influenced by the value of testing pressure (p_1). In Equation 5-39, the denominator of the fourth term of the equation contains these items: fL/D and $2\ln(p_1/p_2)$. The relative magnitude of those two items would determine their influence on LRV calculation. On the one hand, if the value of fL/D is much greater than the value of $2\ln(p_1/p_2)$, then Equation 5-16 can be simplified as follows:

$$\text{LRV} = 1.505 + \log \frac{Q_W \mu_W}{V_H \Delta P \times \text{VCF}} + \log \left(\frac{RT}{M}\right)^{0.5} \frac{(p_1{}^2 - p_2{}^2)^{0.5}}{D(\text{PDR})} - 0.5 \log\left(\frac{L}{fD}\right) \quad \text{(Eq 5-40)}$$

On the other hand, if the value of fL/D is much smaller than the value of $2\ln(p_1/p_2)$, then Equation 5-16 can be simplified as follows:

$$\text{LRV} = 1.505 + \log \frac{Q_W \mu_W}{V_H \Delta P \times \text{VCF}} + \log \left(\frac{RT}{M}\right)^{0.5} \frac{(p_1{}^2 - p_2{}^2)^{0.5}}{D(\text{PDR})} + \log \left\{ \frac{\frac{L}{D}}{\left[2\ln\left(\frac{p_1}{p_2}\right)\right]^{0.5}} \right\}$$

(Eq 5-41)

In the case of Equation 5-40, the same 80 percent change in friction factor causes about one-tenth of a log in LRV. In the case of Equation 5-41, friction factor disappears from the equation altogether. Thus, it does not have an effect on LRV calculation. In current low-pressure systems, the value of $2\ln(p_1/p_2)$ is typically in the range of 1.4 to 2.2, while the range of fL/D can be from a few tenths to a few thousands. Therefore, Equation 5-41 is unlikely to be applicable. Equation 5-40 is applicable to the conditions where hollow-fiber lumen is deemed hydraulically rough; the flow path of the membrane breach is long; and the air flow is in transition regime.

Similar to the interaction between friction factor and the geometry of the membrane breach, testing pressure can also affect the value of the friction factor. As a result, they are not true independent variables. Therefore, sensitivity analysis presented in Table 5-4 should be used with caution. However, as discussed previously, if membrane breach occurs near to the potting interface on filtrate side, the short flow path would diminish the impact of friction factor on LRV. This implies that the resulting error from the use of an inaccurate value of friction factor may not be significant to LRV estimation.

Sensitivity analysis can also provide practical means to improve the sensitivity of detection for membrane breach. For example, a membrane system with a small hold-up volume and high testing pressure can achieve higher sensitivity of pressure-decay test.

VALIDATION OF LRV CALCULATION

Validation of LRV calculation can be performed via a challenge test in conjunction with integrity testing or by directly measuring water bypass flow of known membrane breaches. For a challenge test, a membrane breach can be made by cutting a number of hollow fibers. At different level of integrity breach defined by the number of cut fibers, a challenge test can be performed to obtain data on LRV. At the same time, integrity testing (e.g., pressure-decay test) can also be performed to obtain the parameter indicating for breach (e.g., PDR). Measuring both water bypass flow and air flow can also be performed as an empirical method to determine LRV. The data obtained through the validation test then can be compared with those predicted by the equation relating LRV to integrity testing results (e.g., PDR).

Validate LRV Calculation Via Challenge Test

Ideally, validation of LRV via challenge test can be performed on a scale similar to a production plant. However, it is very difficult to do so as the validation test can be disruptive to the plant production. In this aspect, a research project funded by the American Water Works Association Research Foundation (AwwaRF) provided a rare opportunity for the validation of LRV calculation (Sethi et al. 2004) on the scale of a full-production plant. In this project, microbial challenge test using Bacillus subtilis spores was conducted parallel to seven integrity-testing methods including pressure-hold, turbidity monitoring, multiplex laser turbidity monitoring, particle monitor (using both index and index ratio), particle counter (2 μm and above), and particle amplifier. In the test, a number of hollow fibers in a module were deliberately cut and microbial challenge tests and integrity testing were conducted at different levels of membrane integrity breach and against the baseline measurement (i.e., without membrane breaches). Data on pressure decay test from the research project are presented in this chapter. The data on other integrity testing methods based on water quality monitoring are presented in chapter 6.

Figure 5-7 depicts the LRV and PDR as a function of the number of cut fibers, which indicates that LRV is a logarithmic function of the number of cut fibers (r^2 = 0.99), while PDR is approximately the linear function of the number of cut hollow fibers (r^2 = 0.97). If it is assumed that each cut hollow fiber generated the same bypass flow and air leakage, those relationships are to be expected. In addition, the result proves that the pressure-decay test is very sensitive and able to detect a single cut fiber in a system containing 50 membrane modules (one fiber represents 0.0003 percent of the total hollow fibers for the system).

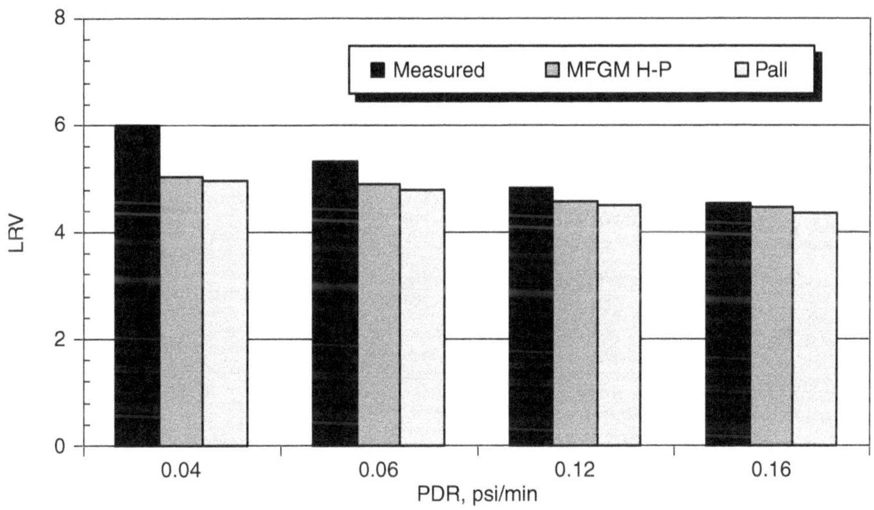

LRV = log reduction value; PDR = pressure-decay rate

Figure 5-8. Comparison of measured LRV and calculated LRV (Equation 5-16) (measured LRV is plotted using data from Sethi et al. 2004)

Figure 5-8 plots LRV as a function of PDR. For comparison, LRV calculated using MFGM H-P Equation and Pall Equation are also plotted on the same graph. Because fibers were cut near the potting on the feed side of the module, it is anticipated that the flow regime in the hollow-fiber lumen is likely laminar due to the long flow path (i.e., large *L/D*). Figure 5-8 shows that the calculated LRV using either equation is always lower than those measured values, and the results of calculated LRV using two equations are very close. The discrepancy between measured and calculated LRV becomes increasingly smaller as the number of cut hollow fibers increases. The result indicates that both equations yield a conservative estimate of LRV. This is expected as a few assumptions used in the development of the equations err on the conservative side. From the regulatory perspective, this kind of conservatism is rather appropriate as it provides a margin of safety.

Validate LRV Calculation Via Measuring Water and Air Flow

A challenge test provides a direct measurement on the impact of membrane breach on filtrate quality. However, conducting a challenge test involves cultivating and measuring microbial reagents or their surrogates, which is complicated and usually time-consuming. As a result, challenge tests involving microbial reagents are usually limited in scope and difficult to conduct under a wide range of conditions. As an alternative, measuring water and air flow through known membrane breaches can provide useful information for validating hydraulic models for LRV calculation because it is simple to perform and can be tested under various conditions. MFGM describes the method to empirically determine ALCR in detail in Appendix D of the MFGM.

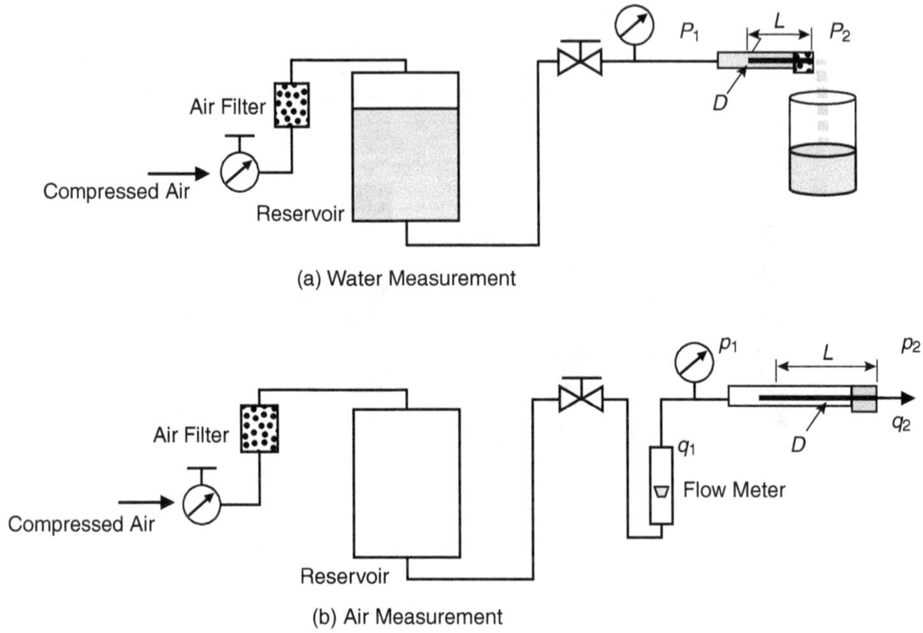

Figure 5-9. Experimental system for measuring water and air flow through a breached hollow fiber

Table 5-6. Variables for the measurements of water and air flow through a breached hollow fiber

Variables	Unit	Values
Water pressure (P_1)	psi	10, 15, 20, 25, 30, 35
Air pressure (p_1)	psi	10, 15, 20, 25, 30, 35
Length of hollow fiber (L)	in. (mm)	2.4 (0.06), 3.9 (0.10), 7.9 (0.20), 13.8 (0.35), 17.7 (0.45), 23.6 (0.60), 31.5 (0.80), 39.4 (1.00), 47.2 (1.20), 59.1 (1.50), 76.8 (1.95)

One example of validating LRV via measuring water and air flow is presented below. Water and air flow through a broken fiber were measured using the system depicted in Figure 5-9. Water and air are under constant pressure in the reservoir and flow through a piece of hollow fiber. The variables chosen for the test include water pressure, air pressure, and the length of fibers, as presented in Table 5-6. At each hollow-fiber length, water pressure was varied and water flow from fiber lumen was measured based on the collected volume in a given time period. The same measurement was repeated for air at various pressures in which air flow was measured by rotameters calibrated for the proper flow ranges.

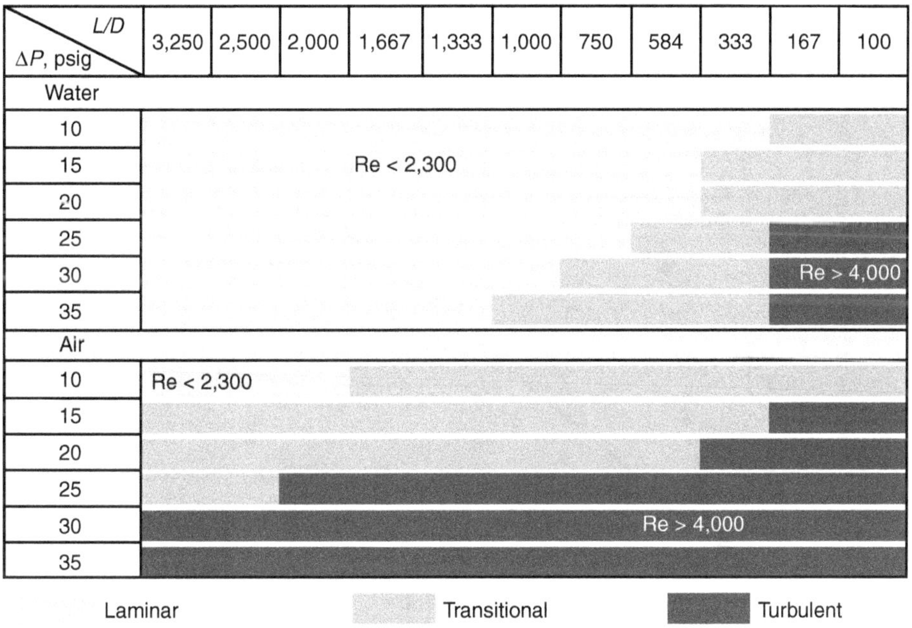

Figure 5-10. Flow regimes under various testing conditions

Flow Regimes of Water Bypass Flow Under Various Testing Conditions

After collecting water and air flow data, the Reynolds number was calculated for a given temperature and diameter of the hollow-fiber lumen to delineate the flow regimes under various testing conditions. The result is presented in Figure 5-10. Two testing variables, pressure differential and *L/D*, are used for constructing 2-D space from which the Reynolds number was calculated. As expected, the laminar flow regime dominates at large *L/D* and low pressure (upper left corner of the panel in Figure 5-10). Conversely, turbulent flow regime dominates at small *L/D* and high pressure (lower right corner of the panel in Figure 5-10). One distinct feature is that the "zones" for each flow regime differ significantly in water flow and air flow. For example, the area for laminar flow regime of water flow is close to two thirds of the total area, while the area for laminar flow regime of air flow occupies less than 5 percent of the space. The area in which water and air flow share the same flow regime (laminar or turbulent) is approximately 14 percent of the total area. Again, they occupy the upper left (laminar) and the lower right (turbulent) corners, respectively.

Discrepancy Between Measured Water-Bypass Flow and Calculated Values of a Completely Broken Hollow Fiber

Because the actual flow regimes may differ from the assumptions used for the calculation of water bypass flow in certain conditions, there are discrepancies in calculated water bypass flow and those actually measured values from a single broken fiber.

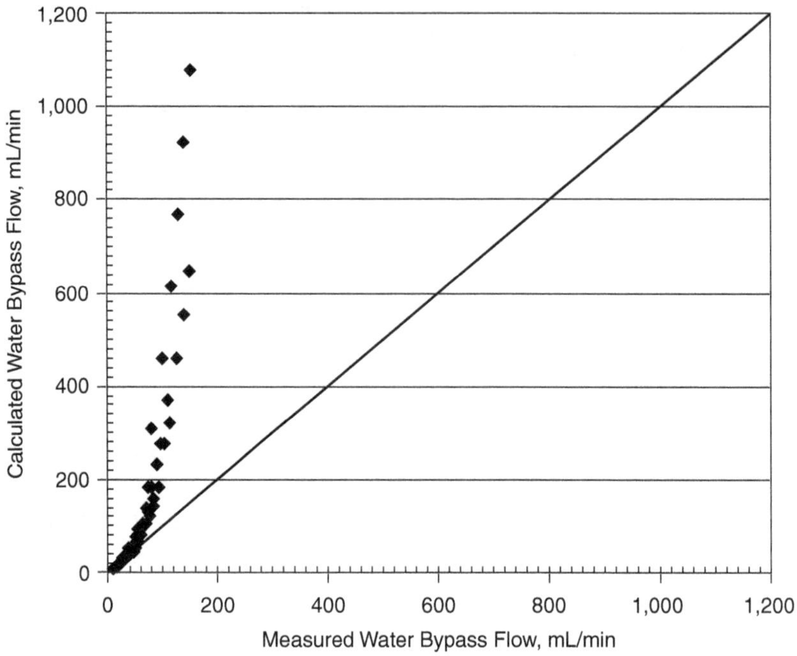

Figure 5-11. Comparison of the measured water bypass flow and those calculated using H-P Equation [*D* (fiber inner diameter) = 0.0006 m; *t* (temperature) = 21°C]

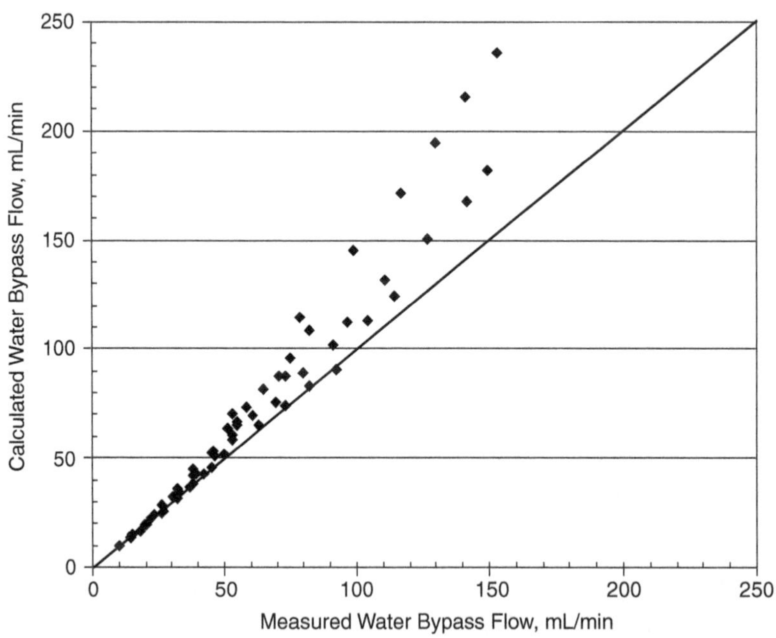

Figure 5-12. Comparison of the measured water bypass flow and those calculated using Darcy Equation [*D* (fiber inner diameter) = 0.0006 m; *t* (temperature) = 21°C]

Figures 5-11 and 5-12 present discrepancies between measured water bypass flow and those based on the calculated values using the H-P Equation and the Darcy Equation, respectively. As indicated in two figures, the majority of the calculated water bypass flow is higher than those of measured values. The discrepancies become more profound as bypass flow increases. When comparing the calculated bypass flows using different equations, it is clear that the Darcy Equation provides a better fit for the experimental data. For example, the discrepancies between measured bypass flow and those calculated using the Darcy Equation ranges from –6 percent to +54 percent as the percent of the measured percent value, whereas it ranges from –15 percent to > +600 percent with the H-P Equation. The difference between the two equations is not surprising because the flow regimes for high bypass flows are more likely out of the laminar regime and a higher margin of error for the calculation is to be expected.

There are several factors that might contribute to the errors from calculation of the bypass flow. The permeation through membranes can have a positive contribution to bypass flow (i.e., increasing bypass flow), and underestimated hydraulic loss due to applying incorrect flow regime and the local losses (i.e., the hydraulic losses occurred at the entrance and exit of the hollow-fiber lumen) in the equations can have a negative contribution to the bypass flow (i.e., reducing bypass flow). Underestimation of bypass flow by the equations due to the omission of permeation through membranes should not affect LRV, because the increased flow is actually filtrate and does not really bypass the membrane. However, overestimation of bypass flow by the equations yields a conservative LRV.

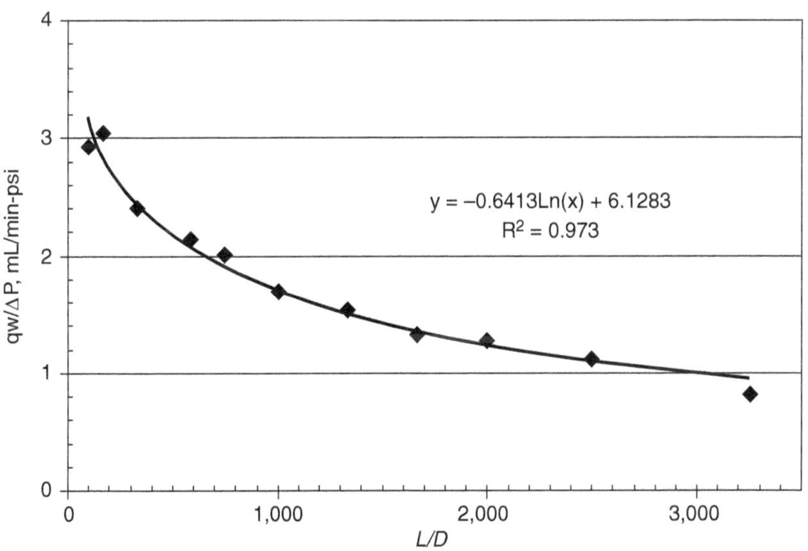

Figure 5-13. Normalized bypass flow as a function of *L/D* [*D* (fiber inner diameter) = 0.0006 m; *t* (temperature) = 21°C]

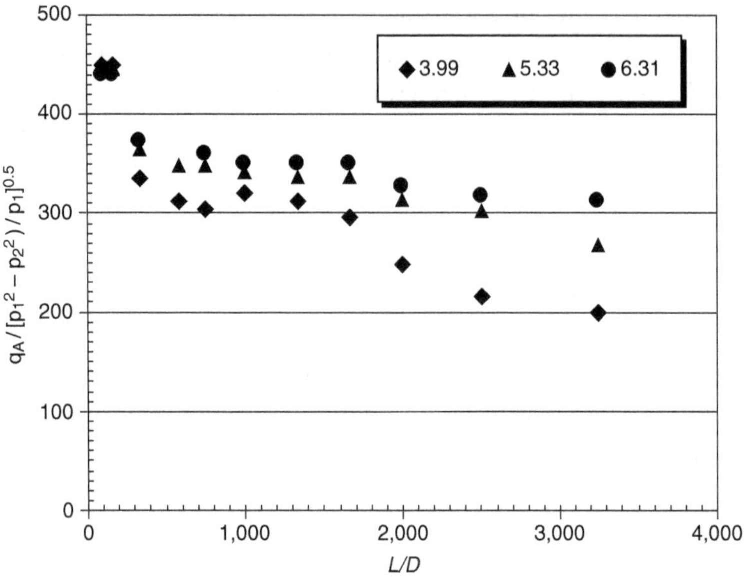

NOTE: The legends are the values of the driving force term ($[(p_1^2 - p_2^2)/p_1]0.5$).

Figure 5-14. Normalized air flow as a function of *L/D* [*D* (fiber inner diameter) = 0.0006 m; *t* (temperature) = 21°C]

A large contribution to the overestimation of bypass flow is likely from the omission of local hydraulic losses for low *L/D* ratio. This is evident in that the calculated bypass flows under the condition of low *L/D* showed the largest discrepancies when compared with the measured values. Figure 5-13 depicts the bypass flow normalized for pressure differential as a function of *L/D*. As indicated in the figure, normalized bypass flow is exclusively a function of *L/D*. Physically, the bypass flow normalized for pressure differential is proportional to the inverse of hydraulic resistance. Therefore, Figure 5-13 indicates that the hydraulic resistance of the hollow-fiber lumen increases as *L/D* increases, following logarithmic function.

Air Flow in the Hollow-Fiber Lumen

Air flow in the hollow-fiber lumen is affected by both *L/D* and pressure differential. According to Equation 5-31, air flow is of hyperbolic form with respect to the driving force,

$$\left(\frac{p_1{}^2 - p_2{}^2}{p_1}\right)^{0.5}.$$

Figure 5-14 depicts the air flow normalized for driving force as a function of *L/D*, in which legends are the values of the driving force term corresponding to a p_1 value of 10, 20, and 30 psig, respectively. The plots in Figure 5-14 have three zones in which the relationship between normalized air flow and *L/D* shows a distinct difference:

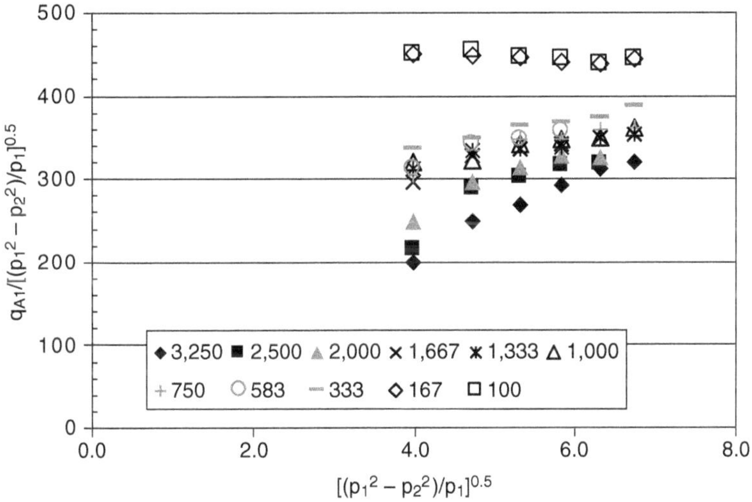

Figure 5-15. Normalized air flow as a function of driving force [*D* (fiber inner diameter) = 0.0006 m; *t* (temperature) = 21°C; the legends are the values of the *L/D*]

1. $L/D \geq 2{,}000$: the normalized air flow decreases as *L/D* increases
2. $333 \leq L/D \leq 1{,}667$: the normalized air flow only decreases very slightly as *L/D* increases
3. $L/D \leq 167$: the normalized air flow is nearly constant

In zones (1) and (2), normalized air flow increases with an increase in driving force, although all plots follow the same trend. In zone (3), air flow converges around the same value, regardless driving force, as well as *L/D* value. Figure 5-15 depicts the normalized air flow as a function of driving force, in which the legends are the values of *L/D*. It shows clearly that the normalized air flows converge around 440–450 mL/min-psi$^{0.5}$ for *L/D* value of 100 and 167, regardless the value of the driving force.

Physically, air flow normalized for the driving force is proportional to the inverse of the resistance. Therefore, data depicted in Figures 5-14 and 5-15 indicate that the resistance to air flow in the hollow-fiber lumen can be considered as a constant when *L/D* falls below certain value. The significance of this finding is that a simple linear equation to describe pressure decay and driving force can be used for calculating LRV, if the membrane breach is known to occur at the point where the flow path in hollow-fiber lumen is near the minimum (i.e., the worst-case scenario).

Calculating ALCR Using the Water Bypass Flow and Air Flow Data

The paired water bypass flow and air flow data can be used to calculate ALCR. An example is given in Figure 5-16 in which the legends represent different values of *L/D*. Figure 5-16 shows that the paired water bypass and air flow have an approximately linear

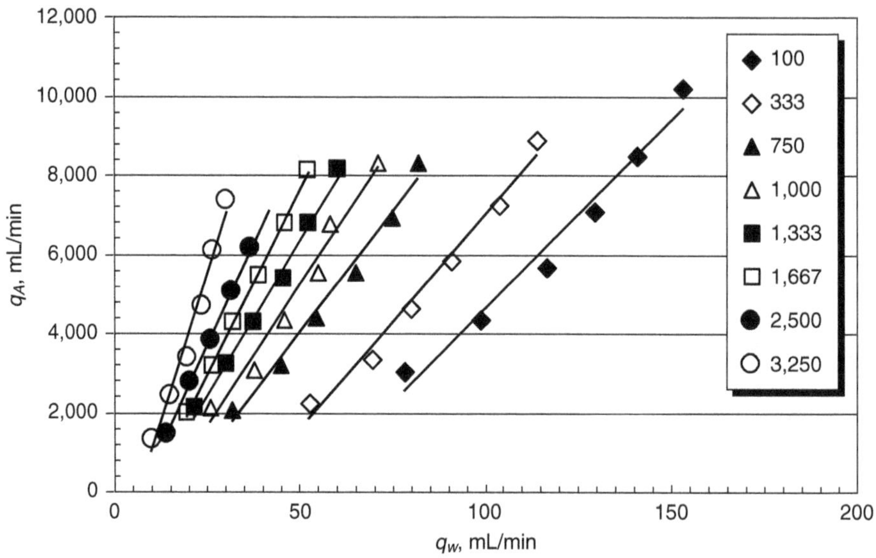

ALCR is the slope of $q_W - q_A$ plot. The legends represent the values of *L/D* .

Figure 5-16. Computing ALCR using paired water bypass flow and air flow [*D* (fiber inner diameter) = 0.0006 m; *t* (temperature) = 21°C]

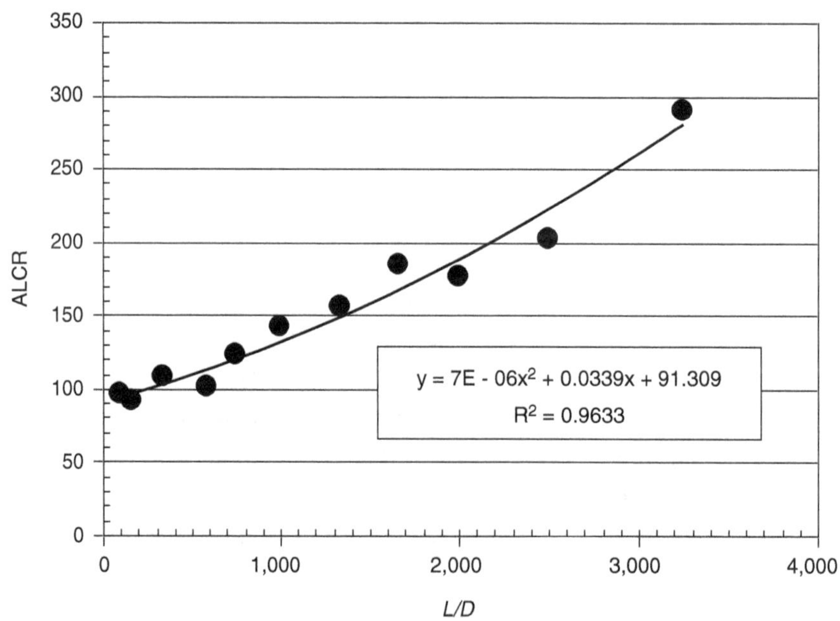

Figure 5-17. The relationship between ALCR and *L/D* [*D* (fiber inner diameter) = 0.0006 m; *t* (temperature) = 21°C]

relationship for each *L/D*. By definition, ALCR is the slope of the water bypass flow–air flow plot. It is clear that ALCR relates to *L/D*: as *L/D* increases, ALCR increases. Figure 5-17 plots ALCR as a function of *L/D*. The dependence of ALCR on *L/D* reflects the asymmetry of fluid dynamics of water and air flow in hollow-fiber lumen: the hydraulic resistance for the air flow in hollow-fiber lumen increases more rapidly than that for water bypass flow as *L/D* decreases.

Although the data in this example are generated with a specific hollow-fiber membrane, the following general observations can be made:

- Both the pressure differential and geometry of the membrane breach (as characterized by *L/D*) affect the fluid dynamics in hollow-fiber lumen. Low pressure differential and high *L/D* tend to have laminar flow, whereas high pressure differential and low *L/D* tend to have turbulent flow.
- The fluid dynamics of water bypass flow and air flow in hollow-fiber lumen can be rather different. For the same pressure differential and *L/D*, water bypass flow tends to have a lower Reynolds number than air flow, simply because the Reynolds number represents the ratio of inertia force to viscous force and water is more viscous than air. As a result, flow regimes of water bypass flow and air flow are the same only under certain conditions.
- Because of the asymmetry in fluid dynamics of water bypass flow and air flow in a hollow-fiber lumen, ALCR should be a function of both the pressure differential and geometry of the membrane breach for hollow-fiber membranes.
- The findings based on the measurement of water bypass flow and air flow can be used to improve the accuracy of LRV calculation.

COMPARISON OF DIFFERENT METHODS FOR LRV CALCULATION

In previous sections, several different methods for LRV are presented. Table 5-7 summarizes the major assumptions and differences of those equations. Because the ASTM method is identical to the H-P Equation in the MFGM, only the H-P Equation in the MFGM is included.

One of the major differences among the three equations is the assumptions of flow regimes for water bypass flow and air flow. The MFGM Darcy Equation assumes both water and air flow in turbulent regimes. This assumption is valid for circumstances of low *L/D* and high pressure differential according to Figure 5-10. However, the MFGM H-P Equation assumes both water bypass and air flows in laminar regimes, which corresponds to the high *L/D* and low pressure differentials according to Figure 5-10. The Pall Equation assumes only water bypass flow in laminar regime. Therefore, it is very conservative for estimating water bypass flow, especially when *L/D* is low and the pressure differential is high. In comparison, water

Table 5-7. Comparison of equations for LRV calculation

Item	MFGM: Darcy	MFGM: H-P	Pall
Conditions and assumptions	• Both water and air flows are turbulent • Resistance coefficients for water and air flows are identical	• Both water and air flows are turbulent	• Water flow is laminar
Basic equations	• Darcy Equations for both water and air flows • Use of expansion factor for compressible fluid (air)	• H-P Equations for both water and air flows • Use of effective testing pressure for compressible fluid (air)	• H-P Equation for water flow • Energy conservation equation for compressible fluid for air
Geometry of membrane breach (*L/D*)	• Implicitly expressed (in the term of expansion factor, *Y*) • ALCR is independent of *L/D* • LRV is dependent of *L/D*	• Not included in the equation • ALCR and LRV are independent of *L/D*	• Explicitly expressed in the equation • LRV is dependent of *L/D*
Result	• The most conservative at high *L/D*, but less conservative than Pall Equation at low *L/D*	• The least conservative	• The most conservative at low *L/D*, but less conservative than the Darcy Equation at high *L/D*

MFGM = *Membrane Filtration Guidance Manual*; H-P = Hagen-Poiseuille equation

bypass flow estimated using Darcy Equation fits better with those actually measured values (Figure 5-12). The Darcy pipe flow equation is a phenomenological equation based on empirical observation but is consistent with the fundamental principles of hydraulics. The equation can be applied to all flow regimes.

Another significant difference among three equations is the role of geometry of membrane breach, as characterized by *L/D*. The MFGM H-P Equation does not include the *L/D* term. This is based on the assumption that equations for water bypass flow and air flow are similar with respect to the geometry of membrane breach. As a result, parameters defining the geometry of the membrane breach are canceled out when relating equations for water bypass flow to that for air flow. Therefore, the MFGM H-P Equation is independent of *L/D*. The MFGM Darcy Equation includes *L/D* in its flow resistance (*K*) term, which in turn embodied in the expansion factor, *Y*. Because the impact of *K* on *Y* is relatively limited, the change in *L/D* does not have a very significant impact on LRV calculation. This insensitive nature of the MFGM Darcy Equation

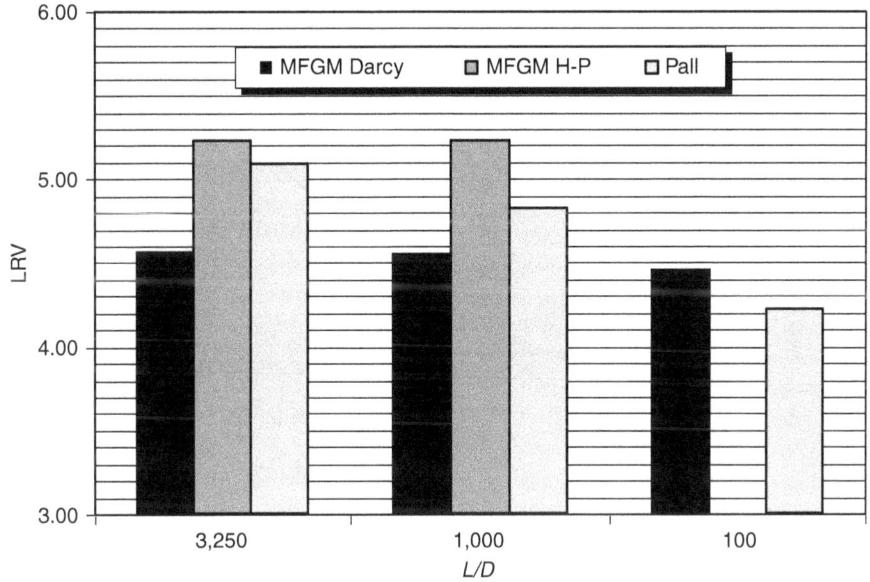

Figure 5-18. Comparison of LRV as a function of *L/D* calculated via the MFGM Darcy, the MFGM H-P, and the Pall Equation. The parameters used in the calculations are: Q_W = 600 gpm; TMP = 10 psi; p_1 (testing pressure for pressure-decay test) = 39.5 psia; *t* = 20°C; PDR = 0.04 psi/min.

can be reflected via the result of sensitivity analysis in Table 5-4, where the change in *L/D* from 100 to 300 only yields a change in resulting LRV of 0.04-logs. In contrast to the MFGM H-P and the MFGM Darcy Equations, the Pall Equation shows a much greater sensitivity to the change in *L/D*. A comparison of LRV calculation using three equations at various *L/D* is depicted in Figure 5-18. Based on the experimental data in Figure 5-10, the flow regimes for water bypass flow are laminar for *L/D* = 3,250 and 1,000, and transitional for *L/D* = 100, respectively. Therefore, LRV calculation for *L/D* = 100 is not included in the graph as the flow regime in this case is not laminar. The results in Figure 5-18 show the following:

- LRV calculated via the MFGM H-P Equation is independent of *L/D*.
- LRV calculated via the MFGM Darcy Equation decreases slightly with decreasing *L/D* (0.1-logs as *L/D* decreases from 3,250 to 100). The MFGM Darcy Equation yields the most conservative LRV at high *L/D* (>1,000), but a less conservative LRV at low *L/D* than that of the Pall Equation.
- LRV calculated by the Pall Equation is the most sensitive to the change in *L/D*. The decrease in LRV is 0.87-logs as *L/D* decreases from 3,250 to 100. This decrease in LRV corresponds to the increase in water bypass flow by a factor of approximately 7.5-fold, matching well to the experimental data (~7.9-fold).

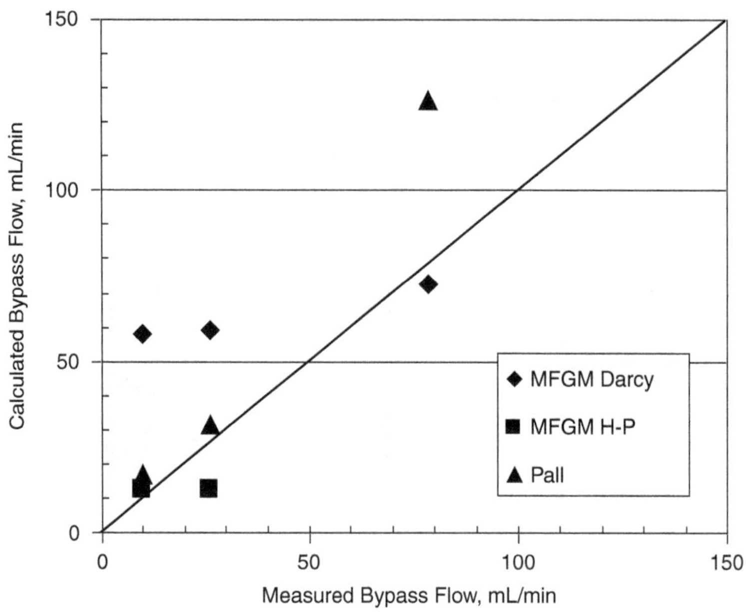

Figure 5-19. Comparison of measured and calculated bypass flow via the MFGM Darcy, the MFGM H-P, and the Pall Equations. The parameters used in the calculations are: *L* = 1.95 m; TMP = 10 psi; *t* = 20°C.

Bypass flows for each equation are calculated and compared with the corresponded measured values as depicted in Figure 5-19. Compared to measured bypass flow, the MFGM Darcy Equation overestimates bypass flow at high *L/D* (1,000 and 3,250) and slightly underestimates bypass flow at low *L/D* (100). However, MFGM H-P Equation has the best estimation for bypass flow at *L/D* = 3,250, but underestimates bypass flow at *L/D* = 1,000. The Pall Equation tends to overestimate bypass flow at all values of *L/D* and therefore provides a conservative result. The discrepancy is more profound at *L/D* = 100, partially due to the increasing departure from the assumption of laminar regime for the bypass flow.

CONCLUSION

In this chapter, membrane integrity testing based on fluid flow is presented. Those testing methods constitute the basis of direct integrity testing as defined in LT2ESWTR. They are the standard integrity testing procedures for the industry because those methods have the advantages of high sensitivity and reliability relative to other methods currently available, and they can be used for tracking down to the individual membrane module(s) that have membrane breaches.

The central issue for membrane integrity testing based on fluid flow is how to relate LRV to the parameters to indicate the membrane breach. This relationship can be established by using either hydraulic modeling or empirical correlation. The hydraulic modeling approach is presented in detail in this chapter. Several equations for LRV calculation are presented. Those equations are developed based on different assumptions. Therefore, one needs to examine if conditions are consistent with the assumptions when applying those equations.

The method to correlate LRV to parameters for indicating membrane breach (namely, air flow or pressure decay) by empirical approach is given detailed coverage in Appendix D of the MFGM, and an example is given in the section on "Validation of LRV Calculation" in this chapter. The validation of the equations for LRV calculation can be performed by either challenge test or measuring of water bypass flow and air flow. Comparing the actual data with predictions from the equations, one can verify whether the assumptions in developing equations are valid. In addition, improvements to the equations can be made by incorporating findings from measured data into the equation.

One important issue is the strong dependence of water bypass flow and air flow on the geometry of a membrane breach defined by *L/D*, as exhibited in Figures 5-13 through 5-18. As *L/D* is not known in practice, for equations that include the term *L/D* (i.e., the MFGM Darcy and the Pall Equations), the current approach has been to assume the worst-case scenario (i.e., assuming the shortest possible flow path from the breach). The argument supporting this approach is that this supposedly yields a conservative estimate for LRV, although the "conservatism" resulting from this approach does depend on the equation being used, as indicated in Figure 5-19. However, in doing so, the parameter indicating membrane breach (e.g., pressure-decay rate) is treated as a de facto variable independent of *L/D*, which is contradictory to the measured data. If the air flow or pressure decay is from the membrane breach of a single hollow fiber, it is possible to solve this problem by correlating the air flow or pressure decay to *L/D*. However, this would not work if multiple hollow fibers break simultaneously with unknown geometry of the membrane breach.

From a regulatory perspective, it may be acceptable if an equation for LRV calculation yields a conservative result so that the water quality of the membrane plant would not compromise pubic health. However, from the operating perspective, equations to predict LRV more accurately are desirable as this would maximize the plant capacity without compromising public health. Therefore, improvements to equations for LRV calculation are still needed based on the better understanding of fluid dynamics of water and air flow through the membrane breach.

Appendix 5-A: Derivation of Equation 5-31

In constructing the mathematic model of airflow through a hollow fiber, the following assumptions are made to simplify the model:

1. Air forms a continuous, steady one-phase flow in a hollow fiber and resembles that in an impermeable pipe. The diffusion of air from and/or into the fiber lumen through fiber walls is negligible.
2. Air behaves like ideal gas.
3. An isothermal condition is assumed in the fiber lumen, i.e., temperature in fiber lumen does not change along the flow path or over time.

The diagram for mathematical deduction is given below. The orientation of the hollow fiber is vertical and air flows upward.

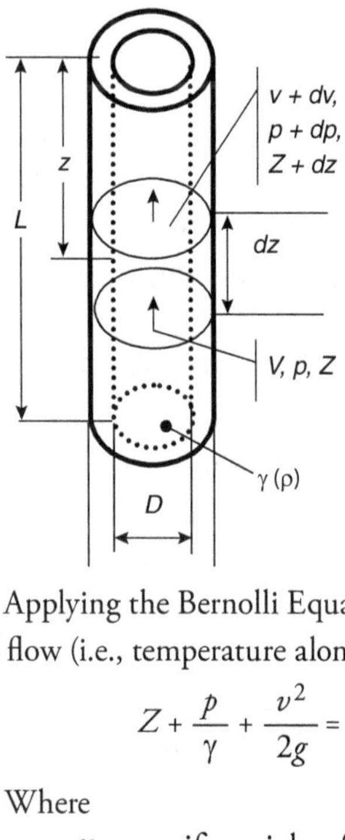

Nomenclatures:

v, p, Z = velocity and pressure inside the hollow fiber, and elevation at the bottom of the fiber segment

$v + dv, p + dp, Z + dz$ = velocity and pressure inside the hollow fiber, and the elevation at the top of the fiber segment

dz = height of the segment of hollow fiber

D = inside and outside diameter of the hollow fiber

L = overall fiber length

γ, ρ = specific weight and density of air, respectively

g = gravity acceleration, 9.81 M/s^2

Applying the Bernolli Equation to the hollow fiber segment and assuming isothermal flow (i.e., temperature along the flow path does not change):

$$Z + \frac{p}{\gamma} + \frac{v^2}{2g} = Z + dZ + \frac{p + dp}{\gamma} + \frac{(v + dv)^2}{2g} + h_f \quad \text{(Eq 5-A-1)}$$

Where

γ = specific weight of air, kN/m^3
h_f = friction loss in the fiber segment, m

Rearrange the equation and neglect the small term including dv^2:

$$dZ + \frac{dp}{\gamma} = -\frac{vdv}{g} h_f \quad \text{(Eq 5-A-2)}$$

For gas, the difference in elevation head, dZ, is much smaller numerically when compared to other terms. Thus, Equation 5-A-2 can be further simplified to:

$$\frac{dp}{\gamma} + \frac{vdv}{g} + h_f = 0 \quad \text{(Eq 5-A-3)}$$

h_f= loss in the fiber lumen in Equation 5-A-3 can be expressed by Darcy Equation:

$$h_f = \frac{fv^2dz}{2Dg} \qquad \text{(Eq 5-A-4)}$$

Where

f = friction factor

Substitute Equation 5-A-4 into Equation 5-A-3 and note by definition $\gamma = \rho g$:

$$\frac{dp}{\rho} + vdv + \frac{fv^2dz}{2D} = 0 \qquad \text{(Eq 5-A-5)}$$

Assuming that air behaves like ideal gas, then

$$pV = \frac{W}{M}RT \qquad \text{(Eq 5-A-6)}$$

Where

V = volume of air, M^3
W = weight of air, kg
M = molecular weight of air, kg/mol
R = gas constant, R = kN-M/°K-mol

By the definition of density, ρ, specific volume, V_S, and specific weight, γ, we have

$$\rho = \frac{W}{V} = \frac{pM}{RT} \qquad \text{(Eq 5-A-7)}$$

$$V_S = \frac{1}{\rho} \qquad \text{(Eq 5-A-8)}$$

Substitute Equation 5-A-7 and Equation 5-A-8 into Equation 5-A-6 and rearrange:

$$V_S = \frac{RT}{pM} \qquad \text{(Eq 5-A-9)}$$

For compressible fluid, Equation 5-A-5 cannot be integrated directly because density of the gas is a variable of pressure. Therefore, the continuity equation based on volumetric flow is not applicable and the continuity equation has to be based on mass flow.

Now introduce the concept of mass flux, U_M,

$$U_M = \nu\rho = \frac{\nu}{V_S} \qquad \text{(Eq 5-A-10)}$$

Thus, $$\nu = U_M V_S \qquad \text{(Eq 5-A-11)}$$

and $$dv = U_M dV_S \qquad \text{(Eq 5-A-12)}$$

Substitute Equations 5-A-8, 5-A-9, 5-A-10, and 5-A-11 into Equation 5-A-5:

$$\frac{M}{RT}pdp + \frac{U_M{}^2 dV_S}{V_S} + \frac{fU_M{}^2 dz}{2D} = 0 \qquad \text{(Eq 5-A-13)}$$

According to continuity equation, the mass flux, U_M, is a constant along the fiber, assuming the gas migrating in and out of the fiber wall is negligible. Thus, integrate Equation 5-A-13 between Point 1 and Point 2 along the fiber length, L, and we have

$$\frac{M}{2RT}(p_2{}^2 - p_1{}^2) + U_M{}^2 \ln \frac{V_{S2}}{V_{S1}} + \frac{fU_M{}^2 L}{2D} = 0 \qquad \text{(Eq 5-A-14)}$$

The subscripts 1 and 2 represent the parameters at the entrance and at the exit of the fiber lumen, respectively.

Rearrange Equation 5-A-14:

$$U_M{}^2 = \frac{1}{\dfrac{f_L}{D} + 2\ln \dfrac{V_{S2}}{V_{S1}}} \times \frac{M}{RT}(p_1{}^2 - p_2{}^2) = \nu_1{}^2 \rho_1{}^2 \qquad \text{(Eq 5-A-15)}$$

$$\text{For ideal gas,} \quad \frac{V_{S2}}{V_{S1}} = \frac{p_1}{p_2} \qquad \text{(Eq 5-A-16)}$$

$$\text{and} \quad \frac{M}{RT} = \frac{\rho_1}{p_2} \qquad \text{(Eq 5-A-17)}$$

Substitute Equations 5-A-16 and 5-A-17 into Equation 5-A-15 and rearrange:

$$\nu_1 = \left[\frac{1}{\rho_1\left(\dfrac{fL}{D} + 2\ln \dfrac{p_1}{p_2}\right)} \; \frac{(p_1{}^2 - p_2{}^2)}{p_1} \right]^{1/2} \qquad \text{(Eq 5-31)}$$

Appendix 5-B: Derivation of Equation 5-35

By definition, the pressure-decay rate (PDR) is given below:

$$\text{PDR} = \frac{P_i - P_f}{t} = \frac{\Delta P}{\Delta t} \qquad \text{(Eq 5-B-1)}$$

Where

P_i = air pressure at the beginning of a pressure-hold test
P_f = air pressure at the end of a pressure-hold test
t = duration of the pressure-hold test

If air behaves like ideal gas, and temperature during the course of testing is constant, PDR can be related to airflow (qA) by the following equation:

$$\text{PDR} = \frac{\Delta n}{\Delta t}\,\frac{RT}{V_H} \qquad \text{(Eq 5-B-2)}$$

Where

$\Delta n/\Delta t$ = molar flow rate of air
R = gas constant, R = 8.314 Pa – M^3/gmol-°K
T = absolute temperature
V_H = air hold-up volume of the membrane system

By definition,

$$\frac{\Delta n}{\Delta t} = q_A\,\frac{\rho^2}{M} \qquad \text{(Eq 5-B-3)}$$

Where

M = molecular weight of air, g/gmol

Substitute Equation 5-B-3 into Equation 5-B-2:

$$\text{PDR} = q_A\,\frac{\rho_2 RT}{MV_H} \qquad \text{(Eq 5-35)}$$

Appendix 5-C: Derivation of Equation 5-39

Air flow through the fiber lumen can be expressed according to Equation 5-34:

$$q_A = \frac{\pi}{4}D^2\left(\frac{p_1}{p_2}\right)\mathrm{n}_1 = \frac{\pi}{4}D^2\left(\frac{p_1}{p_2}\right)\left[\frac{1}{\rho_1\left(\frac{fL}{D} + 2\ln\frac{p_1}{p_2}\right)}\frac{(p_1^2 - p_2^2)}{p_1}\right]^{1/2} \qquad \text{(Eq 5-C-1)}$$

Rearrange Equation 5-C-1:

$$\frac{\pi}{4}D^2 = q_A\left(\frac{p_2}{p_1}\right)\left[\frac{\rho_1 p_1\left(\frac{fL}{D} + 2\ln\frac{p_1}{p_2}\right)}{p_1^2 - p_2^2}\right]^{1/2} \qquad \text{(Eq 5-C-2)}$$

Note q_A can be expressed as a function of PDR per Equation 5-35:

$$q_A = \mathrm{PDR}\frac{MV_H}{\rho_2 RT} \qquad \text{(Eq 5-C-3)}$$

Substitute Equation 5-C-3 into Equation 5-C-2:

$$\frac{\pi}{4}D^2 = (\mathrm{PDR})V_H\left(\frac{M}{\rho_2 RT}\right)\left(\frac{p_2}{p_1}\right)\left[\frac{\rho_1 p_1\left(\frac{fL}{D} + 2\ln\frac{p_1}{p_2}\right)}{p_1^2 - p_2^2}\right]^{0.5} \qquad \text{(Eq 5-C-4)}$$

Note the assumption of isothermal condition in the fiber lumen, thus

$$\rho_2 = \rho_1\frac{p_2}{p_1} \qquad \text{(Eq 5-C-5)}$$

and

$$\frac{p_1}{\rho_1} = \frac{RT}{M} \qquad \text{(Eq 5-C-6)}$$

Substitute Equations 5-C-5 and 5-C-6 into Equation 5-C-4:

$$\frac{\pi}{4}D^2 = (\mathrm{PDR})V_H\left(\frac{M}{RT}\right)^{0.5}\left[\frac{\left(\frac{fL}{D} + 2\ln\frac{p_1}{p_2}\right)}{p_1^2 - p_2^2}\right]^{0.5} \qquad \text{(Eq 5-C-7)}$$

If the bypass flow (q_W) in fiber lumen is in laminar regime during filtration, then it can be expressed as the following:

$$q_W = \frac{\pi}{4}D^2\frac{\Delta PD^2}{32\mu_W L} \qquad \text{(Eq 5-C-8)}$$

Where

ΔP = net pressure differential including static hydraulic head

Rearrange Equation 5-C-8,

$$\frac{\pi}{4} D^2 = \frac{32\mu_W L}{\Delta P D^2} q_W \qquad \text{(Eq 5-C-9)}$$

Combine Equations 5-C-7 and 5-C-9 and rearrange:

$$\frac{1}{q_W} = \left(\frac{32\mu_W L}{D^2 \Delta P}\right)\left(\frac{RT}{M}\right)^{0.5} \frac{1}{(\mathrm{PDR}) V_H} \left[\frac{p_1^2 - p_2^2}{\left(\frac{fL}{D} + 2\ln\frac{p_1}{p_2}\right)}\right]^{0.5} \qquad \text{(Eq 5-C-10)}$$

If microbial pathogens are completely removed by a filter, as the case of *Giardia* and *Cryptosporidium* by microfiltration of submicron rating filters, the LRV can be expressed as following:

$$\mathrm{LRV} = \log \frac{Q_W}{\mathrm{VCF} \times q_w} \qquad \text{(Eq 5-C-11)}$$

Where

Q_W = filtrate flow

Substitute Equation 5-C-10 into Equation 5-C-11:

$$\mathrm{LRV} = 1.505 + \log \frac{Q_W \mu_W}{\mathrm{VCF} \times V_H \Delta P} + \log \left(\frac{RT}{M}\right)^{0.5} \frac{(p_1^2 - p_2^2)^{0.5}}{D(\mathrm{PDR})} + \log \left\{ \frac{\frac{L}{D}}{\left[\frac{fL}{D} + 2\ln\left(\frac{p_1}{p_2}\right)\right]^{0.5}} \right\}$$

(Eq 5-16)

6

MEMBRANE INTEGRITY TESTING BASED ON MONITORING FILTRATE QUALITY

INTRODUCTION

As discussed previously, the purpose of membrane integrity testing is to ensure the safety of the finished water; that is, to make sure the finished water is pathogen free. Direct measurement of waterborne pathogens is, of course, desirable. Unfortunately, the measurement of these pathogens is often very complicated and time-consuming. From a practical point of view, direct measurement of waterborne pathogens has not advanced to the point where it can be used to monitor membrane integrity. Therefore, integrity testing based on water quality monitoring for low-pressure membrane systems is typically performed on surrogates for waterborne pathogens. The most common and basic methods are turbidity monitoring and particle counts. Several other methods are extensions of those two methods and require changes in hardware (e.g., light source from visible light to laser), in signal measurement and data processing (e.g., particle monitor based on particle count index instead of the counts themselves), and in the way measuring is undertaken (e.g., multiplex sensors and marker). The result is improved resolution, sensitivity, and reliability of the measurements.

For this reason, turbidity and particle counts are discussed in-depth first and then the discussion is extended to other methods derived from those two. The methods include laser turbidity (i.e., laser nephelometer), particle monitor, multiple sensing, marker-based membrane integrity monitoring, and particle amplifier. The discussion of integrity-monitoring method begins with a description of the working principle of the method. Factors affecting the measurement are then discussed, followed by the application of the method to monitor membrane integrity.

Actual data are used to illustrate the responses of various monitoring methods to an integrity breach. A major source of data is the Awwa Research Foundation (AwwaRF) report *Assessment and Development of Low-Pressure Membrane Integrity Monitoring Tools* (Sethi et al. 2004), which provides the basis for many of the graphs in this chapter. At the end of the chapter, the various methods based on water quality monitoring are compared and their feasibility for monitoring membrane integrity is discussed.

TURBIDITY

Turbidity is a visual characteristic of fluid. The word is used to refer to the cloudy appearance of a water sample. Arguably, the measurement of turbidity may relate to the quantity of suspended matter in water. However, the value of turbidity depends on the way it is measured. The current standard measurement for turbidity in the United States is the nephelometric turbidity unit (ntu)—the turbidity measured by an instrument called a nephelometer.

A nephelometer consists of a light source and a light detector set on one side of the source beam. Usually, the detector is set perpendicular to the direction of the light beam (i.e., a scattering angle of 90 degrees). When the light beam shines on the water sample, light is scattered by the suspended matter in the sample. The light detector picks up the light reflected on it and sends the signal to the instrument's circuitry, which compares the signal to the internal standard. The signal is then calibrated using formazin polymer and generates a reading. To some extent, the intensity of the light reflected on the detector depends not only on the quantity of the suspended particles but also on the size, shape, and optical property (reflectivity) of the particles and the matrix effect of water.

Impact of Particle Size

The intensity of light scattering is strongly related to the ratio of particle size to the wavelength of the light. For small particles (≤1/10 of the wavelength of the light), light scattering is isotropic (evenly distributed in all directions). As the size of particles increases, light scattering is increasingly concentrated in the forward direction. Because the light detector in a standard nephelometer is placed to capture light at a 90-degree scattering angle to the light beam, large particles would generate a lower turbidity reading than small particles on the basis of equal mass.

Figure 6-1 illustrates how the size of spherical particles affects specific turbidity—turbidity normalized on the volume of particles occupied per unit volume of suspension (Gregory 2006). The light source in a standard nephelometer generates white light —a spectrum of visible lights with a wavelength range approximately from 380–750 nm. (A typical tungsten lamp has a spectrum extended well into the infrared range, but a light source filter reduces much of the infrared radiation reaching the sample.) As a result, particles smaller than 0.05 μm in diameter scatter light almost equally in all directions while particles above 1 μm scatter light in the forward direction.

Note that the impact of particle size in Figure 6-1 is based on single scattering condition. In the actual turbidity measurement, multiple scattering often occurs. The light reflected by a particle can scatter again when it meets another particle. With this condition, the interaction tends to be averaged out by the large number of scattering events, and the overall result tends to be more random. In other words, the impact of particle size

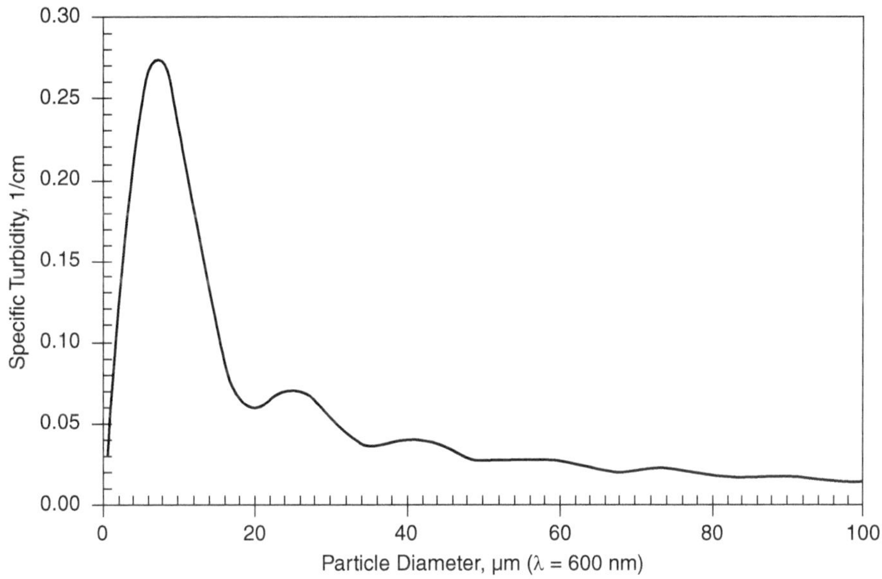

Figure 6-1. Specific turbidity as a function of the size of spherical particles

is lessened, and the trend illustrated in Figure 6-1 becomes more flattened. Nevertheless, particle size has still significant impact on the result of turbidity measurement. This has important implications for monitoring particles in water because the measurement is not sensitive to detect particles larger than a few µm in size (Gregory 2006).

Impact of Reflectivity of Particles

Reflectivity is the fraction of incident light reflected by a surface. The higher the reflectivity of a material, the higher the intensity of the light it scatters. Reflectivity is a function of the ratio of the refractive indices of the particles to the medium in which particles exist at a given wavelength of the light. The refractive index of water at 20°C is 1.33 (λ = 589.3 nm). The more a particle differs from water in refractive index, the more light-reflecting the particle is. A material that does not reflect light at all is called *blackbody* and has zero reflectivity. Thus, a particle of powdered activated carbon (PAC) would reflect light poorly. As a result, PAC particles in water produce a lower turbidity reading than mineral particles that are of the same size and shape but have higher reflectivity.

Biological matter such as bacteria or algae also has lower reflectivity than minerals. Figure 6-2 compares the plots of specific turbidity with particle diameters of mineral and biological matter. As indicated in the figure, specific turbidity is generally lower for biological matter than for minerals. Size-dependence, as well, is much more diminished for biological matter than for minerals.

It should be noted that reflectivity depends on the direction of measurement. Therefore, it is a directional property. According to the direction of the light reflection,

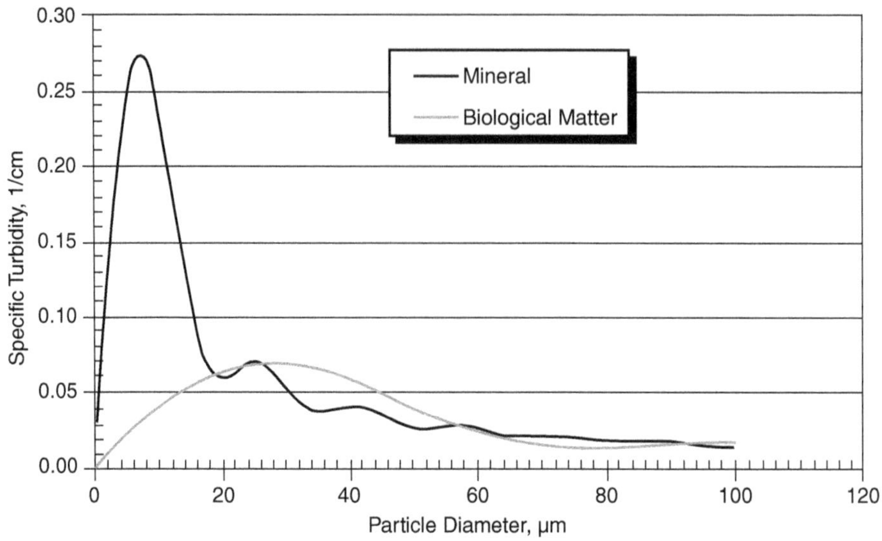

Figure 6-2. Comparison of specific turbidity with particle sizes of minerals and biological matter (wavelength λ = 600 nm)

surfaces can be classified as *specular* or *diffuse*. For a specular surface, light is reflected only at the appropriate angle and at all other angles reflectivity is near zero. A diffuse surface is the opposite. It reflects light almost equally at all angles and so reflectivity is nearly uniform across all angles. A good example of the former is glass or polished metal; for the latter, matte white pane. Specular and diffuse surfaces are two extremes of ideal surfaces. Surfaces of indigenous particles in water are somewhere in between.

Impact of the Color of Water Matrix

Water gets its color by absorbing certain spectra of sunlight (white light). Because the light source of a standard nephelometer is also white, water with color can absorb the light from the nephelometer. This results in reduced intensity of incident light. Consequently, it would cause a lower turbidity reading as the intensity of scattering light is also reduced. Many color-inducing substances are actually colloids of organic matter—products of decomposition of vegetation and algae. Therefore, as long as the light-absorption spectra of the water sample overlap with the incident light from the light source of the instrument, the reading would generate a negative bias for turbidity (i.e., a lower turbidity reading) as an indicator of suspended matter (Sadar 2002a).

Impact of Particle Concentration

The relationship between the intensity of light scattering and particle concentration is not always linear. At low particle concentration, the intensity of light scattering increases linearly with increasing particle concentration. As particle concentration

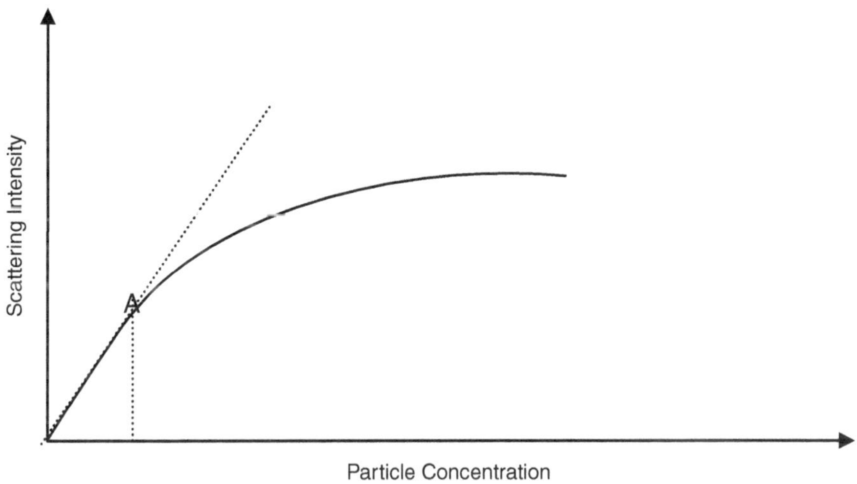

Figure 6-3. Intensity of scattering light as a function of particle concentration

increases further, scattering intensity starts to flatten out until it reaches a plateau. The trend of scattering intensity as a function of particle concentration is illustrated in Figure 6-3.

As particle concentration increases beyond point A in Figure 6-3, the linear relationship between scattering intensity and particle concentration is no longer valid. Therefore, the particle concentration corresponding to point A is the upper limit for accurate turbidity readings. Increased secondary light scattering caused by increasing particle concentration appears to contribute to the departure of the linear relationship. The secondary scattering reflects light out of the periphery of the light detector. Consequently, the overall scattering intensity captured by the light detector decreases. The impact of secondary scattering is illustrated in Figure 6-4.

Using Nephelometric Turbidimeter to Monitor Membrane Integrity

The use of turbidity as a surrogate for microbial quality of finished water has a long history. Turbidity is regulated in the Surface Water Treatment Rule (USEPA 1989). In the Long-Term 2 Enhanced Surface Water Treatment Rule (LT2ESWTR), turbidity monitoring is designated as the default method for indirect integrity monitoring for membrane processes. The major issues of turbidity monitoring are its low sensitivity and reliability. The filtrate turbidity of low-pressure membrane processes is almost always <0.1 ntu. Turbidity does not increase even when membranes have a minor integrity breach.

Figure 6-5 provides an example. In this example, a hollow-fiber microfiltration (MF) module on a 50-module rack had a fixed number of hollow fibers cut and turbidity was monitored as a function of the number of cut fibers. The feedwater turbidity varied between 1 and 2.5 ntu with one exception, when it peaked to 5 ntu during the

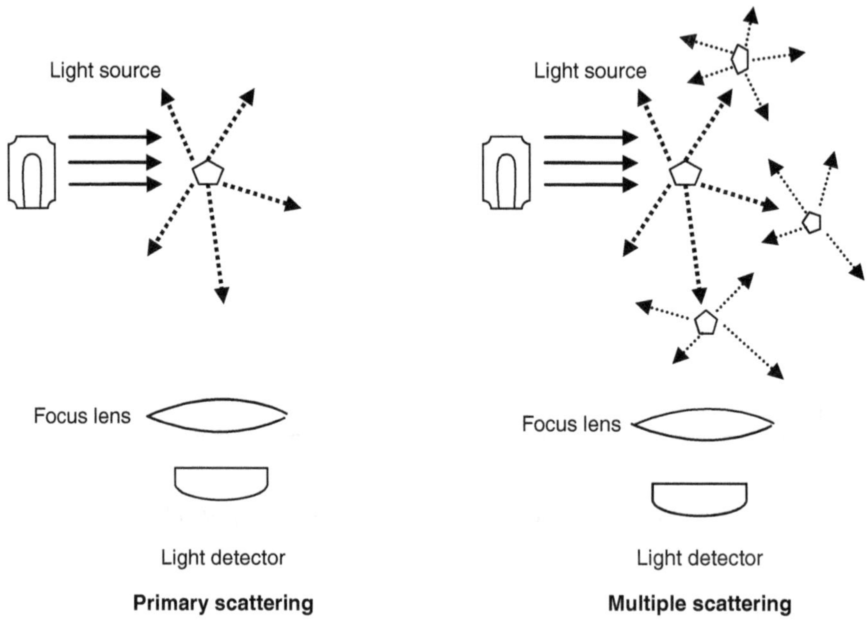

Figure 6-4. Impact of increasing secondary light scattering on measuring turbidity

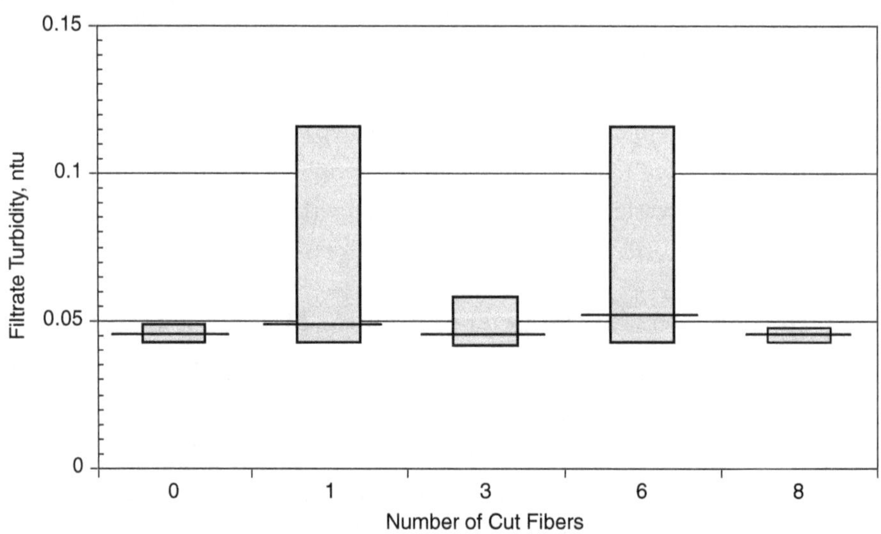

Figure 6-5. Turbidity responses to membrane integrity breaches. The box-and-whisker plot indicates minimum, mean, and maximum values, respectively (plotted from data of Sethi et al. 2004).

test for six cut fibers. The box-and-whisker plots show the maximum, minimum, and mean value of filtrate turbidity as a function of the number of cut fibers. As indicated in Figure 6-5, means of filtrate turbidity fluctuated and did not correspond to the degree of membrane breach (i.e., the number of cut fibers).

It should be noted that absolutely pure, particle-free water scatters light as a result of thermal fluctuations in refractive index, giving an effective turbidity of 0.03 ntu (Gregory 2006). The minimum value of turbidity reading from MF filtrate for all cases in Figure 6-5 is approximately 0.04 ntu—very close to the background "turbidity" caused by light scattering of the pure water. Therefore, it is not surprising that nephelometer could not distinguish whether the membranes are integral.

Another reason turbidity monitoring is inappropriate as a tool for monitoring membrane integrity is the large dilution of the bypass flow through the defect by the filtrate from intact membranes. As shown in Figure 6-5, even in the event that eight fibers were cut, the dilution factor (the ratio of filtrate flow from intact membranes to bypass flow from cut fibers) was still more than 10,000. However, the standard nephelometer is only capable of accurately measuring down to 0.3 ntu (Sadar 2004). Therefore, the change in turbidity at such a large dilution simply could not be detected by a turbidimeter.

Based on the evidence, turbidity monitoring is inadequate to detect small membrane integrity breach due to the monitoring's lack of sensitivity and reliability. Its usefulness as an indicator of membrane integrity is likely to be limited to small membrane systems filtering high-turbidity water.

LIQUID PARTICLE COUNTERS

Types of Liquid Particle Counters

Particle counts are the number of particles in a fixed volume, typically in 1 mL of fluid. The result for particle counts is generally expressed as number of particles per mL. The instrument used for particle counts is called a *particle counter*. Based on how the measurement is made, there are three types of particle counters: light-blocking counters, light-scattering counters, and Coulter counters. The first two are optical particle counters based on light blocking and light scattering, respectively; and the last one is a mechanical/electrical device based on the change in impedance (resistance to electrical current) caused by the passing of particles through an orifice placed between a pair of electrodes.

Coulter counters are mostly used in medical fields, such as for counting red blood cells. Those counters work best when the sample fluid contains high concentration of electrolytes and the particles to be measured are nonconducting particles. Because water typically contains a low concentration of electrolytes and particles in water typically carry some electrical charges, the use of Coulter counters is less feasible and will not be discussed in this book.

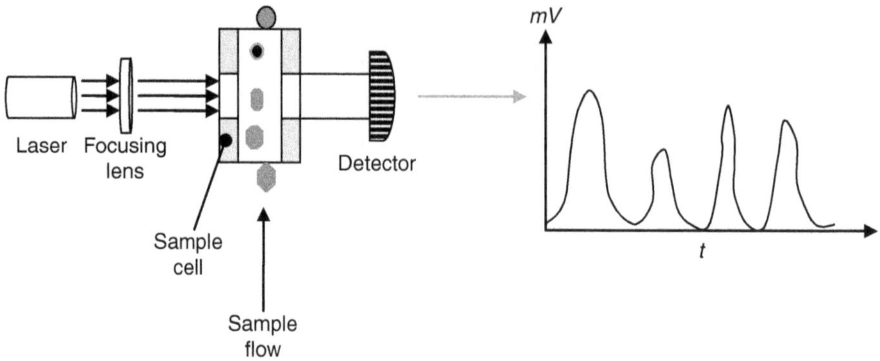

mV = millivolts; *t* = time

Figure 6-6. Schematic of light-blocking optical particle counter

Light-Blocking Particle Counter

The light-blocking type of optical particle counter is typically used to detect and size particles greater than 1 µm. Most particle counters used in the drinking water industry are of this type. The working principle of the light-blocking counter is illustrated in Figure 6-6. A water sample flows through a sample cell that has windows to allow laser light passing through. The particles intersecting the light path block light, and the light detector generates an electrical voltage pulse. The amplitude of the signal is proportional to the amount of the light blocked by the particle. Thus, the amplitude of the signal is correlated to the size of the particle.

The particle size is determined by setting the upper and/or lower threshold values for the amplitude of the signals. The upper and lower threshold values are calibrated with a standard for each range of particle sizes, or channel. Any signal that has amplitude either greater than the upper threshold or smaller than the lower threshold is not be counted for the size range. Therefore, particle counts can be sorted for different size ranges. The particle counters most commonly used in the drinking water industry typically have multiple size ranges (channels) that count particles from 2 µm to several hundreds of micrometers. However, because large particles tend to settle, there is little use for counting particles over 100 µm. Most counters are set for four channels: 2–5 µm, 5–10 µm, 10–15 µm, and >2 µm. Note that *Cryptosporidium* oocysts and *Giardia* cysts have size ranges of 3–5 µm and 5–8 µm, respectively, which fall right into the size range of counters.

Even though a particle counter sorts particles according to size, it should be noted that optical particle counters do not really measure size, but the light intensity received on light detectors. That is, a particle counter actually measures light attenuation or light extinction. Therefore, light loss such as that from scattering and absorption by particles is interpreted as being blocked by the particle. From this perspective, the optical property of a particle does affect the measurement of its size.

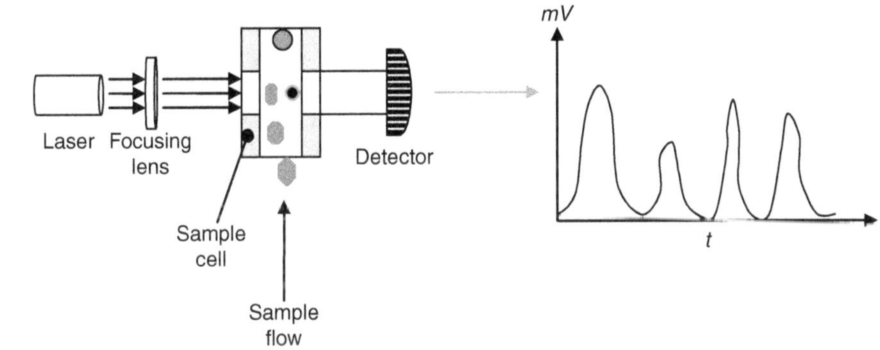

a. Undercount caused by one particle obscuring the other

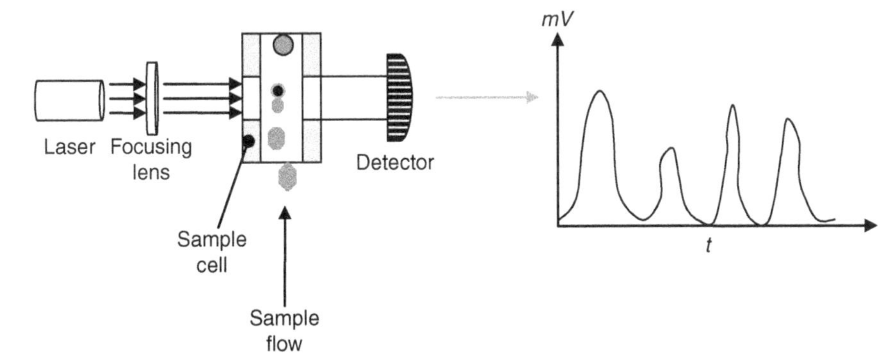

b. Oversizing caused by two particles passing head to tail

mV = millivolts; *t* = time

Figure 6-7. Coincidence of optical particle counters

As discussed in the section on turbidity measurement, large particles (1 µm and above) scatter light mostly in a forward direction. This can generate a negative bias and result in underestimates of actual particle size. However, isotropic light scattering by small particles or light absorption by the water matrix can reduce the overall light intensity received by the light detector and consequently generates a positive bias for size measurement.

Particle concentration can affect the accuracy of the count. At high concentration, light-blocking particle counters are likely to undercount the particles and/or to overestimate the size of particles—a phenomenon called *coincidence*. Coincidence occurs when two or more particles pass the light path not in sequential mode (single file) but side-by-side. This phenomenon is illustrated in Figure 6-7. All particle counters have an upper concentration limit, typically in the range of 10,000–30,000 particles/mL. This upper concentration is typically expressed as <10 percent coincidence loss at the given concentration limit—that is, an error in particle count of <10 percent.

Light-Scattering Particle Counter

A light-scattering particle counter measures light scattered by the particles that are passing the light path in the sample cell. As with a nephelometer, the light detector is typically placed at an angle of 90 degrees to the light path. Unlike a nephelometer, the light source used in the particle counter is a laser. Thus, it has a well-defined wavelength and very little light diverging along the path—very useful characteristics when it comes to particle size and particle counts.

The major advantage of the light-scattering particle counter is its ability to size and count smaller particles that a light-blocking counter cannot count. Typically, this technique enables the counting of particles down to 0.05 μm or larger. Therefore, a light-scattering particle counter is considered a high-sensitivity counter. (It should be noted that the words *sensitivity* and *resolution*, used when discussing particle counters, have different meanings than in chapter 3. In this chapter, *sensitivity* refers to the smallest particle size that a counter can size and count; *resolution* refers to the ability of a counter to differentiate sizes of particles.)

As in the discussion for nephelometer, the factors affecting light scattering all influence the sizing and counting of particles. Light scattering by a particle is related to its size. If a particle has a size about one tenth of the wavelength of the light or smaller, light scattering is isotropic. Light scattering in this region is termed *Rayleigh scattering* and light-scattering intensity in this region is a function of the sixth power of the particle size. The Rayleigh region is near the lower threshold of high-sensitivity light-scattering particle counters. For counters measuring particle size ≥0.1 μm, light scattering becomes increasingly less size dependent.

The use of a high-sensitivity particle counter can be rationalized because it enhances the sensitivity of integrity monitoring, based on the fact that the size distribution of aquatic suspension typically follows power law (Stumm and Morgan 1996):

$$\frac{dN}{dd_p} = Ad_p^{-\beta} \qquad \text{(Eq 6-1)}$$

Where

d_p = the particle diameter (assuming sphere shape)
N = the number of particles with a diameter of d_p
A and β = coefficients determining shape of size distribution curve

It has been suggested that $\beta = 4$ when colloids are produced by erosion (Lerman 1979). A plot of particle size distribution based on Equation 6-1 is presented in Figure 6-8. As indicated by the plot, small particles outnumber the large ones by a wide margin. For example, the ratio of 0.2 μm particles to 2 μm particles is 1,000:1. Data for a tap water sample measured by a high-resolution counter seem to confirm the validity of the equation, as illustrated in Figure 6-9. Very few particles have a size above

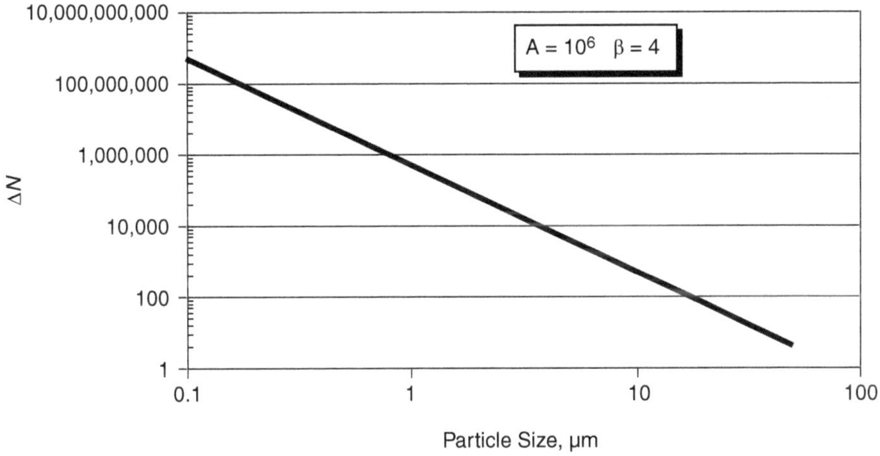

Figure 6-8. Hypothetical particle size distribution in aquatic suspension (calculated from Equation 6-1)

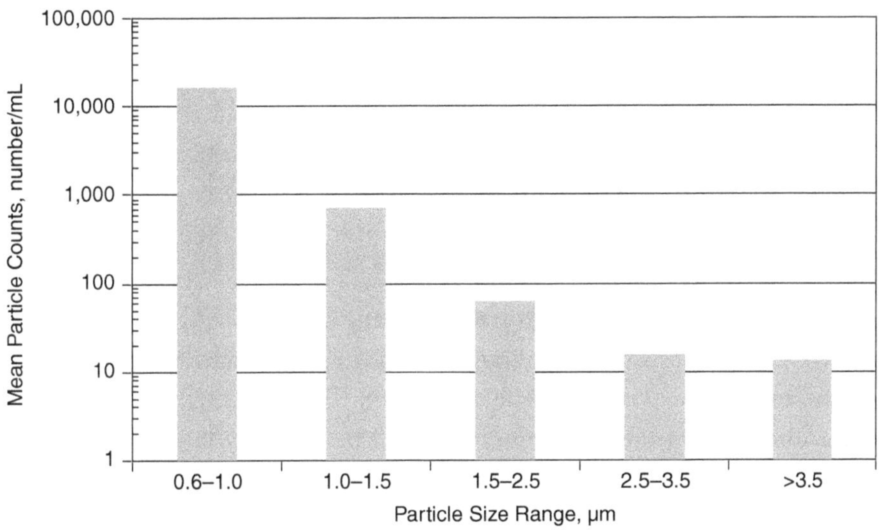

Figure 6-9. Particle size distribution measured on a tap water sample

3.5 μm (15 /mL on average), unlike particles in the size range of 0.6–1.0 μm (approximately 17,000/mL on average)—a difference of more than three orders of magnitude.

Using Particle Counters to Monitor Membrane Integrity

The use of particle counters in water treatment is relatively recent (Bellamy et al. 1993). It has been suggested that a particle counter is more sensitive to the change in performance of granular media filters (Cleasby and Logsdon 1999). This is particularly the case for water with low turbidity (e.g., <0.1 ntu), for which standard turbidity meters are near the limit of detection.

One advantage of using particle counters is the ability to provide information about particle sizes. Conceptually, this should allow the operator to monitor the process with a method capable of meeting resolution criterion as specified in the LT2ESWTR (USEPA 2006a), although the rule does not require resolution criterion to be met for indirect integrity monitoring. However, it is rather difficult to do so in reality, because of the sensitivity and reliability issues of particle counters used for low-pressure membranes.

In addition to the factors affecting counts caused by the inherent physical principles used by optical particle counter discussed previously, many other factors can also affect accuracy of counts.

- **Limitation of the instrument.** A perfect instrument would show a zero count when there were no particles larger than the lower sensitivity limit in a water sample. However, it is not unusual for a particle counter to show a low reading even though, in fact, no particles above the sensitivity limit exist. Several factors may contribute to this. Stray light and current leakage of the circuit are two often-cited reasons (Sadar 2002a). Generally, noises from those sources are random.
- **Particle shedding.** A low-pressure membrane system is not a particle-free system. The large surface area of piping, moving parts such as pumps and valves, and variations in flow velocity and direction all can create conditions that cause shedding of particles — either those attached to the surface, or those embedded in system materials. Particle shedding can also be very random and can contribute to an increased noise level.
- **Dilution effect of low-pressure membrane systems.** As discussed in chapter 5, the low-pressure membrane system is characterized by large dilution to the bypass flow from a membrane defect. This implies that particle counts resulted from a membrane breach would be generally low in comparison to the noise level. That is, conditions under which the optical particle counters are operated create a typical low signal-to-noise ratio, which in turn causes problems in measurement sensitivity and reliability.
- **Influence of feedwater quality.** Particle counts in the filtrate of membranes are directly related to those in the feed stream. Low particle counts in the feed stream (e.g., in the range of a few thousands per milliliter) would be unable to generate a signal significantly above the noise level when a membrane system had a low-level integrity breach. Therefore, particle counts cannot be used as a sensitive and reliable tool for membrane integrity monitoring. In addition, water quality can change from time to time. It is difficult to set control limits if the change in water quality of the feed stream is considerable.
- **Interference from the operation.** If the water is saturated with air, micron-sized bubbles can form in the filtrate as pressure on the filtrate side becomes

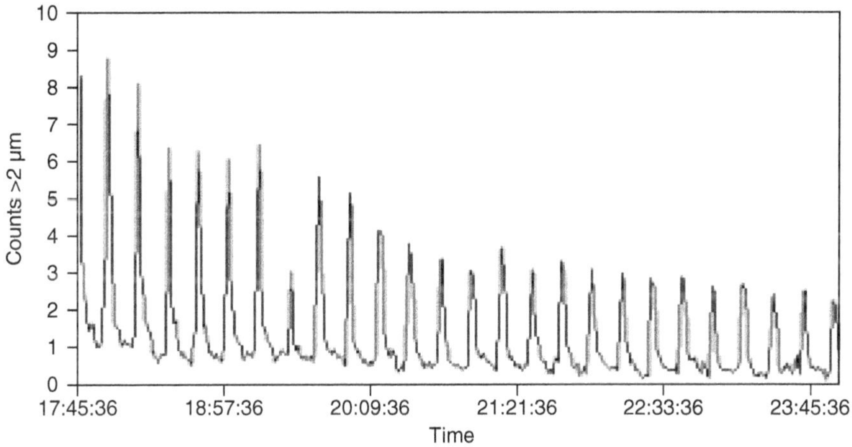

Figure 6-10. Periodical peaking of counts caused by air-assisted backwash

lower than the pressure on the feed side. Those bubbles would be registered by a particle counter, as they too, can either block or scatter light. In addition, some membrane systems apply air-assisted backwash. This operation can result in counts that peak periodically, as illustrated in Figure 6-10, even though the particle counter was equipped with bubble traps in this example in Figure 6-10.

The relation of particle counts to membrane integrity breach is illustrated in Figure 6-11. The box-and-whisker plot represents minimum, average, and maximum values of particle counts for each case of integrity breach (the number of cut fibers). Again, the plot shows that average counts do not correlate to the degree of membrane integrity breach and that there is a large overlap in particle counts between the intact membrane (zero cut fibers) and the compromised membranes. Therefore, the particle counter was not only unable to indicate if a membrane system was integral, but it also failed to differentiate the degree of membrane integrity breach.

It is intriguing to see that the intact membrane (i.e., zero cut fibers) still had particles greater than 2 µm (ranging from ~0.1 to 8 /mL) in its filtrate. The membranes were integrity tested by a pressure-decay test under a pressure of 30 psi; this corresponded to a pore diameter of approximately 1.4 µm. Therefore, no particles greater than 2 µm should have been in the filtrate. The sources discussed previously may have contributed to the high noise level of particle counts. The high noise level of particle counts, combined with large dilution effect typical of low-pressure membrane systems, results in a very low signal-to-noise ratio. This is the primary reason that particle counters are neither sensitive enough nor reliable enough for detection of low-level membrane integrity breaches.

Low particle counts in the filtrate stream of low-pressure membrane systems make it very difficult to identify low-level membrane integrity breaches, as any differences

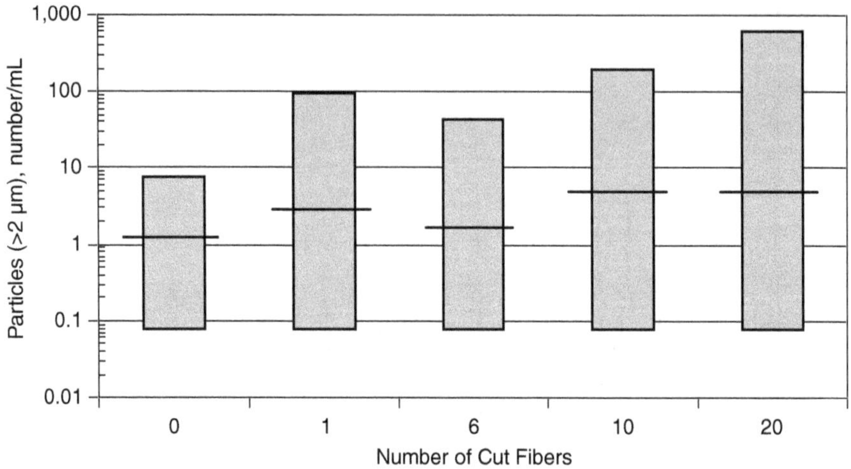

Figure 6-11. Responses of particle counts to membrane integrity breaches. The box-and-whisker plot indicates minimum, mean, and maximum, respectively (plotted from data of Sethi et al. 2004).

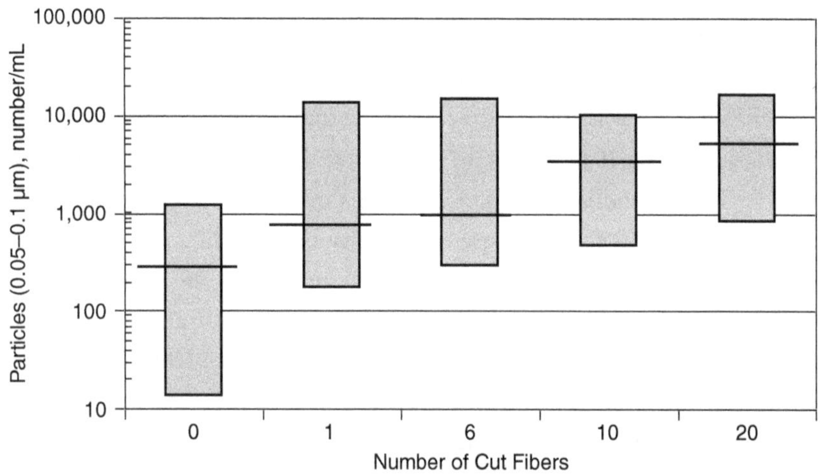

Figure 6-12. Responses of high-sensitivity particle counts to membrane integrity breaches. The box-and-whisker plot indicates minimum, mean, and maximum, respectively (plotted from data of Sethi et al. 2004).

in counts are likely statistically insignificant. As discussed earlier, the small particles outnumber large ones in natural water. Therefore, the use of high-sensitivity particle counters may improve the situation. Figure 6-12 shows particle counts of the same set of tests as in Figure 6-11 but measured with a high-sensitivity counter capable of measuring particles ranging in size from 0.05 to 0.1 µm. The particle counter readings in Figures 6-11 and 6-12 were from the same rack that had multiple membrane modules containing about one third of a million hollow fibers. Figure 6-12 shows the significant improvement in both testing sensitivity and reliability.

First of all, the average counts correlate with the degree of membrane integrity breach, although the counts still were not proportional to the number of cut fibers. Second, the spread (the range of minimum and maximum counts) also becomes narrower—generally within two orders of magnitude, versus more than three orders of magnitude in Figure 6-11. Nevertheless, the large overlap of count readings indicates how difficult it can be to use particle counters (even high-sensitivity counters) to reliably monitor such levels of membrane integrity breach. However, for a smaller membrane system with a higher degree of membrane breach, or for a raw water with a higher particle concentration, particle counter may be used for monitoring membrane integrity breach, providing the control limit is properly established.

LASER NEPHELOMETER

Comparison to Standard Nephelometer

Laser nephelometers are similar to regular nephelometers, with several significant improvements. Besides changing the light source to laser, which has a very narrow wavelength range (around 660 nm), a laser nephelometer collimates the incident light into a narrow beam. This results in a high beam power density and enhances the laser nephelometer's sensitivity by increasing the scattering intensity from small particles. The signal from the light receiver is carried by fiber optics transmission to the light detector, which reduces noise level and produces a very low and stable baseline. In addition, a light trap placed at the far end of the light beam absorbs the light passing through the water sample without scattering it, to further reduce the stray light. These improvements are reported to greatly enhance the sensitivity and reliability of turbidity measurement (Sadar 2004).

Overall, the laser nephelometer has a sensitivity of two orders of magnitude over the standard nephelometer. As a result, the laser nephelometer measures turbidity by mntu, or one thousandth of ntu. Sadar (2004) reported that the detection limit (defined as the minimum detectable change in turbidity) of the laser nephelometer is down to 0.3 mntu. This would represent an improvement of nearly three orders of magnitude over the standard nephelometer. The limit of quantification (LOQ) for the laser nephelometer is about 10 times of detection limit, giving a value about 3 mntu (Sadar 2006).

Using Laser Nephelometer to Monitor Membrane Integrity

Figure 6-13 presents the laser nephelometer responses to membrane integrity breaches in the same fiber cutting test as was used in plotting Figure 6-5. Data from the laser nephelometer indeed showed an improved sensitivity compared with data in Figure 6-5. First of all, the laser nephelometer showed a much lower filtrate turbidity range from 0.2 to 15 mntu in Figure 6-13 versus 0.042 to 0.116 ntu (42 to 116 mntu) in

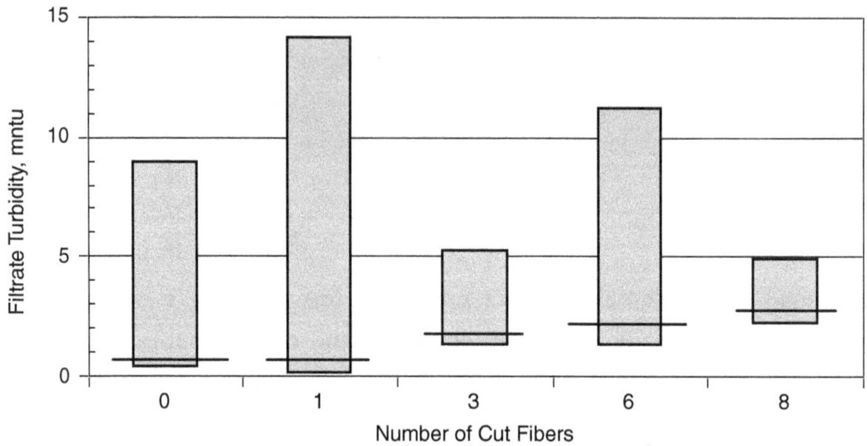

Figure 6-13. Responses of laser nephelometer to membrane integrity breaches. The box-and-whisker plot indicates minimum, mean, and maximum, respectively (plotted from data of Sethi et al. 2004).

Figure 6-5, indicating an enhanced detection limit. Second, the average filtrate turbidity measured by the standard nephelometer did not correlate the degree of membrane breach at all. Measured by the laser nephelometer, however, that correlation exists to a large extent.

One exception is that for the laser nephelometer, the baseline reading (zero cut fiber) is slightly greater than the reading with one cut fiber, indicating that a low reading is beyond the detection limit of the instrument. However, the enhanced detection limit did nothing to increase the reliability of the measurement, which is illustrated by the large spread between the minimum and maximum values (i.e., the height of the boxes) in Figure 6-13. In fact, the enhanced detection limit actually did the opposite. For example, the ratio of maximum value to minimum value of filtrate turbidity ranges from 2.2 to 74 for the laser nephelometer, compared with 1.1 to 2.7 for the standard nephelometer. This example proves that merely increasing detection sensitivity does not necessarily guarantee the quality of the measurement.

PARTICLE MONITOR

At first glace, a particle monitor is similar to a particle counter in configuration: It has a light source, a sample cell, a light detector, and circuits to process the signal. However, the working principle of a particle monitor is not quite the same as that for a particle counter. A particle monitor uses an infrared light-emitting diode (LED) with a wavelength of 850 nm. Such an instrument is suitable to measure particles greater than 1 μm and insensitive to the particles below 0.5 μm (Chemtrac Systems 2003). In contrast to the particle counters, a particle monitor does not measure light

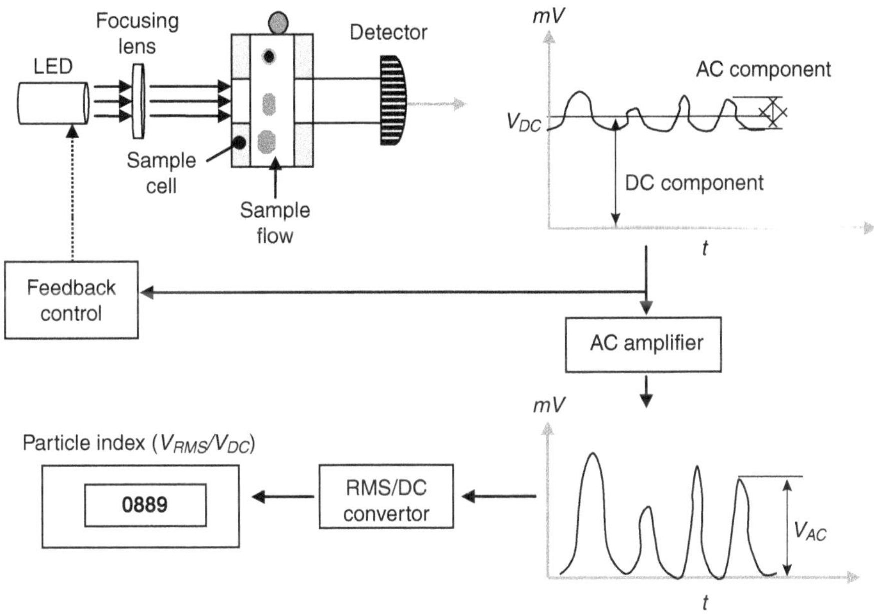

NOTE: LED = light-emitting diode; AC = alternating current; DC = direct current; RMS = root mean square

Figure 6-14. Schematic diagram of working principle for particle monitor

blocking or light scattering per se, but the relative changes in light intensity. The schematic diagram of the particle monitor is illustrated in Figure 6-14.

The signal from the detector consists of two components: a large and steady component (V_{DC}) representing the average light intensity, and a small and fluctuating component (V_{AC}) caused by the change of the light intensity caused by the passage of particles in the light beam. The amplitude of the alternating current (AC) component is a function of both the size and the number of particles at any given time. The direct current (DC) component is kept constant by feedback control, while the AC component is separated from the DC component, amplified, and converted into a DC voltage (V_{RMS}) that is equal to the root mean square (RMS) value of the amplified AC signal. That is,

$$V_{RMS} = \sqrt{\frac{1}{N}\sum_{i=1}^{N}(V_{AC}{}^{2}i)} \qquad \text{(Eq 6-2)}$$

Or using the continuous form of the signal (with respect to time t),

$$V_{RMS} = \sqrt{\frac{1}{T_2 - T_1}\left(\int_{T_1}^{T_2} V_{AC}(t)^2 dt\right)} \qquad \text{(Eq 6-3)}$$

The RMS signal is then recombined with the DC component to generate a ratio:

$$R = \frac{V_{RMS}}{V_{DC}} \quad \text{(Eq 6-4)}$$

This ratio, R, is defined as particle index.

Particle index does not quantify the amount of particles in the water sample, nor does it provide sizing information as particle counters do. However, it provides a relative measurement of the impairment of water quality expressed on an arbitrary scale from 0 to 9,999 (a lower value indicates better water quality). The advantage of this method is the ability to eliminate the drift of baseline value caused by the deterioration of circuit stability or fouling of the sample cell, because those would affect V_{RMS} and V_{DC} in the same proportion and would cancel out in the ratio measurement expressed in Equation 6-4. As the particle monitor does not quantify the number of particles, the need for elaborate calibration of the type used for particle counters can also be eliminated. The value of R depends on the amplification of the AC component. Consequently, its absolute value is not relevant. What is significant is the change in R when comparing a measurement with the baseline value (i.e., the integral membranes).

For a polydispersed particle population, the value of R can be related to particle concentration, N_i, and the light scattering cross-section, a_i, by the following equation (Gregory 1989):

$$R \propto \left(\sum_i N_i a_i^2\right)^{0.5} \quad \text{(Eq 6-5)}$$

According to Equation 6-5, particles with a large light-scattering cross-section, a_i, make the most significant contribution to R. Therefore, a relatively small number of large particles would change R more significantly than a much larger number of small particles. This, in principle, seems very feasible to monitor the membrane integrity where a few large particles are likely to be present when there is a membrane breach.

Using Particle Monitor to Monitor Membrane Integrity

The response of the particle monitor to membrane integrity breach is presented in Figure 6-15 as a box-and-whisker plot of the filtrate particle index (minimum, average, and maximum values). It shows that even though the average value of the filtrate particle index increased significantly when a membrane integrity breach occurred, the relationship between the filtrate particle index and the degree of the breach was not always monotonic. In addition, the large overlap of filtrate particle index values across the test range prevents using the filtrate particle index as a reliable indicator for breaches in membrane integrity.

One problem with using the filtrate particle index is that the parameter is very sensitive to the change in feedwater quality, to the extent that the change in the filtrate particle index overshadows the difference in filtrate particle index values caused by

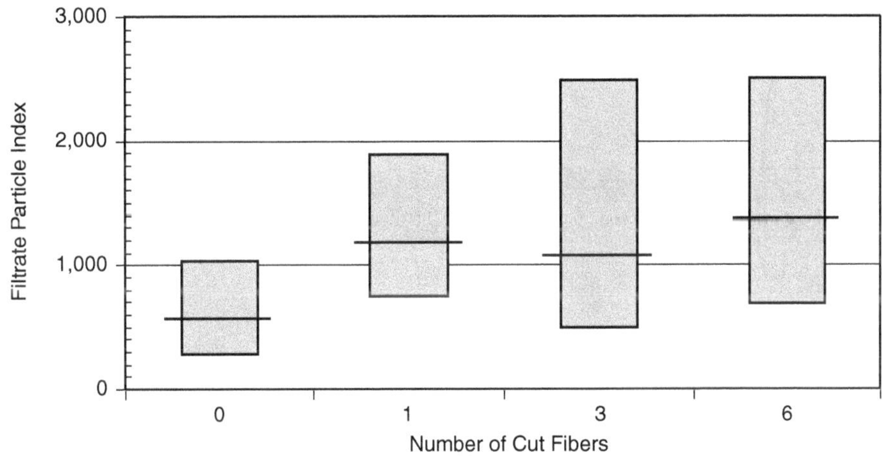

Figure 6-15. Particle monitor responses to membrane integrity breaches. The box-and-whisker plot indicates minimum, mean, and maximum, respectively (plotted from data of Sethi et al. 2004).

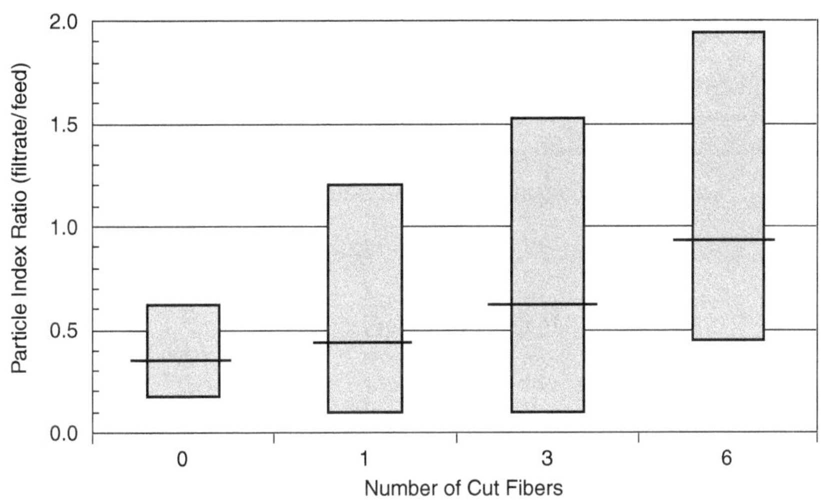

Figure 6-16. The ratio of filtrate-to-feed particle indices as a function of membrane breaches. The box-and-whisker plot indicates minimum, mean, and maximum, respectively (plotted from data of Sethi et al. 2004).

the different degrees of membrane breaches. One way to address the influence of the variations in feedwater quality is to use the ratio of the particle indices for filtrate and feed to normalize. Figure 6-16 presents the result of the ratio of the particle indices as a function of the number of cut fibers. As indicated in Figure 6-16, normalization does improve the correlation between the average particle index ratio and the degree of membrane breaches. However, it does not improve the reliability of the measurement as indicated by the overlap of particle index values.

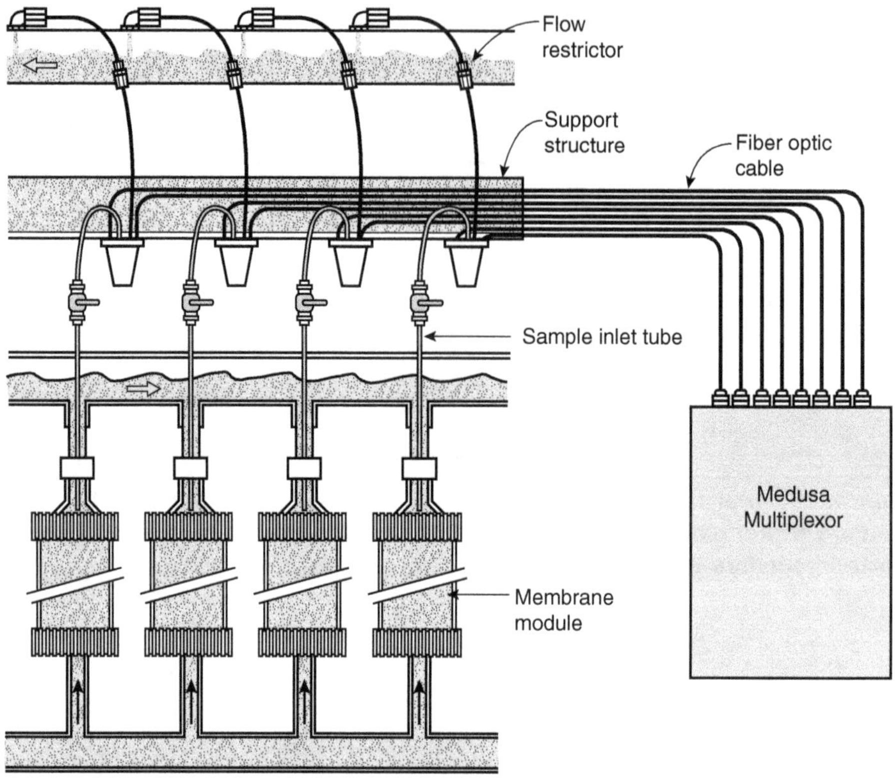

Figure 6-17. Schematic diagram of a multiple-sensing system (Sadar et al. 2003a)

MULTIPLE-SENSING MONITORING

From previous discussions, it is obvious that none of the methods based on monitoring filtrate quality is sensitive and reliable enough to detect a small breach in membrane integrity, such as a few broken hollow fibers on a rack consisting of multiple membrane modules. A major reason for this is the large dilution factor characteristic of low-pressure membrane systems that generates a very low signal-to-noise ratio. Apparently, if the signal-to-noise ratio can be increased, the sensitivity and reliability of monitoring can be improved. Therefore, it is natural to wonder what would happen if each membrane module were equipped with a sensor.

That is the idea for a setup called *multiplex monitoring*. As the name implies, multiplex monitoring uses a sensor array on a multi-module rack. The system uses a shared light source and optical and electrical components to reduce cost. Figure 6-17 illustrates this concept. The entire system consists of several components: the sensors; the fiber optic light transmitting cables; and a base station consisting of a multiplexor (a high-accuracy mechanical positioning system) and shared optical and electrical components to generate, detect, and process signals. The multiplexor sends incident light

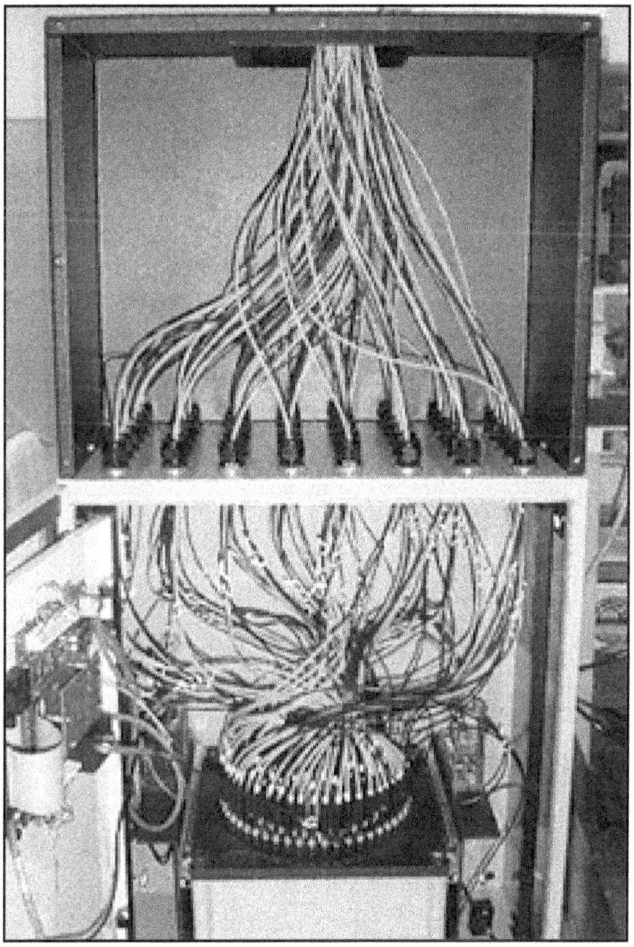

Figure 6-18. Multiplexor and fiber optic cable connections for a 48-sensor array (Sadar et al. 2003b)

to and retrieves the signals from the sensor array via fiber optic cables in a sequential manner. That is, the incident light is first directed to the first sensor in the array and the resulting signal from it is processed first, followed by the signals from the second sensor, the third sensor, and so on.

A photo of such a multiplexor is presented in Figure 6-18. The system uses a statistical method to process two types of data: the mean (e.g., light scattering) and the relative standard deviation (RSD), which is the value of the standard deviation divided by the mean of a set of consecutive measurements (typically 10 consecutive measurements). Therefore, RSD is the same as the coefficient of variation (CV). The RSD measurement is based on the theory that when membrane integrity is breached, the passing of large particles (e.g., 1 to 3 μm) would cause considerable variation in the signal strength over the length of a measuring period. Therefore, it provides a more sensitive way to monitor membrane integrity.

Because of the difference in response of individual sensors, signals from each sensor need to be calibrated and normalized to set the control point to indicate membrane breach. Typically, the control point is set at three standard deviations (3 σ) above the mean to ensure a confidence level of 99.5 percent (Sadar et al. 2003c). In addition, the size of the sensor array may also be limited to ensure that one measuring cycle (i.e., the time needed to complete the measurement of all the sensors in an array) can be completed within 15 minutes—the longest time interval that can be considered to be continuous monitoring as defined in the LT2ESWTR.

If the processing time for the signal from a sensor is Δt (in seconds), the time to complete one cycle of monitoring on N sensors is equal to $N\Delta t$. The maximum number of the sensors in an array, N_{MAX}, is calculated by:

$$N_{MAX} = \frac{900s}{\Delta t} \quad \text{(Eq 6-6)}$$

For most applications, the sensor array is set up for one sensor per membrane module. This not only would enhance the sensitivity of the measurement but also would provide a way to trace the individual module with the integrity breach.

Using Multiple-Sensing to Monitor Membrane Integrity

To date, the actual applications of multiple-sensor array are still very few. However, the preliminary data seem encouraging. In one such application, a laser turbidity sensor is used as a measuring device, and both the mean value (known as the *calc parameter*) and the relative standard deviation (RSD) for 10 or more consecutive measurements were used to indicate membrane integrity. Figures 6-19 and 6-20 present the results of a fiber cutting/pinning test from four sensors mounted on a module. As indicated in those graphs, in spite of variations in the individual sensors' responses to the membrane integrity breach, overall they showed a distinct correlation between sensors' responses (the calc and RSD values) and the degree of the membrane breach. Table 6-1 summarizes the results of the same data set (Sadar 2003). The data indicate that the mean values (calc) of the measurement did not change significantly when membrane was breached, while RSD values were significantly higher than the baseline (zero fiber cut). However, the correlation between RSD value and number of breached fibers is rather poor. As a result, this method may be able to signal membrane breach, but it is not likely to provide a quantitative measurement to indicate the severity of membrane breach.

MARKER-BASED INTEGRITY MONITORING

As discussed in previous sections, a major focus for integrity testing based on water quality is to increase the sensitivity of the measurement by increasing the strength of the signal. This approach is not always helpful if enhanced sensitivity also elevates the

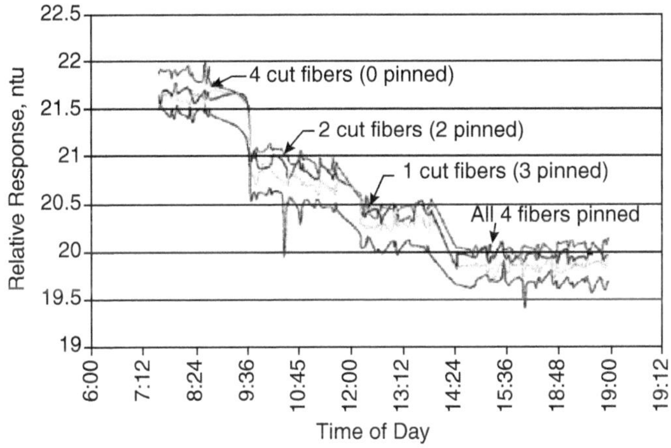

Figure 6-19. Multiple-sensing responses (Calc) to membrane integrity breach (Sadar et al. 2003a)

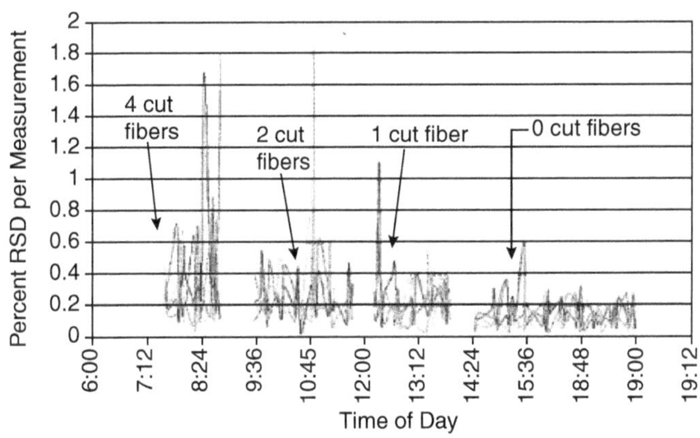

Figure 6-20. Responses of multiple sensing (relative standard deviation, or RSD) to membrane integrity breach (Sadar et al. 2003a)

Table 6-1. Summary of results from multiple-sensing test (Sadar 2003)

Number of Cut Fibers	Calc*	RSD†	Percent Change‡ in Calc (%)	Percent Change‡ in RSD (%)
0	20.25	0.15	0.0	0.0
1	23.68	1.00	16.6	400.7
2	27.77	1.55	36.4	672.1
4	24.18	0.75	19.4	406.3
7	22.27	1.36	22.3	812.4
14	22.97	1.76	13.4	1084.4

* Calc = mean
† RSD = relative standard deviation
‡ The baseline values (0 cut fibers) as 0%

noise level of the measurement. One common issue for all integrity testing methods based on monitoring water quality is the variation in quality of the incoming raw water. This variation in raw water quality can change the baseline and/or diminish the signal strength resulting from membrane breach. Consequently, the result may not be reliable even though method itself is rather sensitive. Another approach to solve this problem is to enhance the sensitivity of the measurement while damping noise at the same time. Marker-based integrity monitoring can be viewed as such an approach.

Description of the Method

Marker-based integrity monitoring involves injecting a marker onto the feedwater upstream of the membrane and monitoring the marker in the membrane filtrate. The sensitivity of this method can be directly calculated by measuring the marker both upstream and downstream of the membrane.

Selection of the marker is the most critical element of this method. The marker can be either indigenous to raw water or treatment processes, or it can be a foreign substance that is neither harmful to human health nor detrimental to treatment processes in any manner. Even though the *Membrane Filtration Guidance Manual* (USEPA 2005) states only that the marker has to meet the performance criteria for the LT2ESWTR and "must be inert and suitable for use in a water treatment facility," regulatory authorities are expected to scrutinize any markers that are foreign. In addition, adding something into the water can also be objectionable to the customers consuming the water, regardless of whether the marker has health effects.

Using Marker-Based Method to Monitor Membrane Integrity

An example of marker-based integrity monitoring is *spiked integrity monitoring* (SIM). This method of spiking powdered activated carbon (PAC) into the feed stream has been evaluated for an ultrafiltration plant both at pilot scale and at a full-scale plant (Franklin, Smith, and Knops 2001; Van Hoof, Kruithof, and Kamp 2001). A specific grade of PAC that has a mean size less than that of *Cryptosporidium* and is reported to have no effect on membranes is used as the marker. PAC slurry is continuously stirred to keep the particles in suspension. After injection, particle counters located upstream and downstream of the membranes are used to monitor the counts. Then the log reduction value (LRV) of the particles is calculated. Figure 6-21 presents the results of a pilot test in which particle counts with and without spiked PAC are compared.

As indicated in the figure, PAC spiking increased particle counts in the feed by approximately a factor of 20. This could damp the change in particle counts in the feedwater. Furthermore, when the counts in the filtrate with and without membrane breach are compared, PAC spiking resulted in an increase in counts of nearly 6-fold over the baseline (i.e., without membrane breach), whereas the control (without PAC

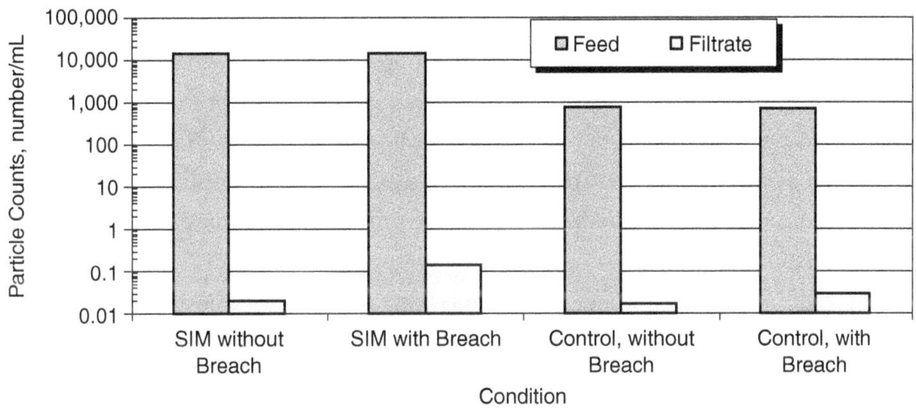

SIM = spiked integrity monitoring

Figure 6-21. SIM test results (plotted based on data of Van Hoof, Kruithof, and Kamp 2001)

spiking) showed an increase in counts of merely 70 percent over the baseline. Therefore, SIM provides a far more sensitive and reliable indicator for membrane integrity breach.

As in the previous example of the spiked integrity monitoring, the monitoring of the marker relies on particle counts. Therefore, the accuracy and reliability of counters contribute to the success or failure of the method (Franklin, Smith, and Knops 2001). A major operational issue for marker-based integrity monitoring is to maintain marker injection and monitoring systems (e.g., particle counters) in good working order. In some cases, this requires a considerable amount of the operator's time. In addition, continuous marker-injection may not be practical. For typical operation, the tests are scheduled on a periodic basis via a programmable logic controller (PLC) (Sethi et al. 2004).

EVALUATION OF DIFFERENT METHODS BASED ON MONITORING WATER QUALITY

The objective of membrane integrity testing is to ensure that the membrane system is integral so that the quality of the filtered water is not compromised for human consumption. In that sense, water quality monitoring is really the end, not the means. However, once the previous discussion is examined to see what is actually monitored in the practice, two parameters emerge: turbidity and particles (and some derivatives of the two). Note that they are merely the surrogates that indicate possibly compromised filtered water quality and do not necessarily have health impact of their own. Then, the questions become:

1. Can monitoring of those surrogates of water quality be definitive indicators for membrane integrity?
2. To what degree can those methods be used and what are their limitations?

Essentially, those are the same issues as discussed in chapters 3 and 5.

As all the methods discussed in the chapter 6 employ on-line instrumentation, they are considered to be continuous monitoring —with very few exceptions. Therefore, testing frequency is not an issue. However, with the exception of the multiple-sensing method, none of these methods is capable of tracing the integrity breach to the individual module, so that the issue of traceability is not applicable. As a result, sensitivity and reliability of the methods are the only relevant issues and thus the focus of the discussion.

Sensitivity is about whether a membrane system is integral and to what degree the integrity of the membranes is compromised. Reliability is about confidence in the answers. Sensitivity and reliability are so highly intertwined in practice that it is difficult, if not impossible, to discuss them separately. If the result of integrity monitoring does not have a high confidence level, it may be useless for the operator to take an action.

Sensitivity

Sensitivity can be evaluated by comparing the values of monitored water quality parameters with varying degrees of membrane breach to the values of an integral membrane system. The change in values can signify a membrane breach. Because of the random nature of the measurement, a statistical approach is the proper way to give some measure of confidence to the result. Assuming that the measurements follow normal distribution and an increased value above the baseline (integral membrane), to give a 95 percent confidence level is deemed to give a statistically significant result. Then the sensitivity of a method can be determined by examining at what degree of membrane breach, as represented by the percentage of broken fibers, the measured value would meet the criterion. In other words, if the average value of a measured parameter for the breached membrane is greater than the value of the baseline (integral membrane) at a confidence level of 95 percent, a breach is deemed to be detected, and the sensitivity of detection is presented as the percent of cut fibers based on the given membrane systems.

The results from an AwwaRF research project (Sethi et al. 2004) for eight methods based on water quality monitoring across several membrane systems are summarized in Table 6-2.

Several observations are made in the AwwaRF report:

- The sensitivity of methods varied from one membrane system to another. This is not surprising considering that differences in membrane configuration and operation are considerable from system to system.
- In general, particle counters seems more sensitive than other methods. For the standard particle counters (> 2 μm) among the six membrane systems tested, three showed a sensitivity of 0.001 percent cut fibers, whereas for the other three systems, the counters were unable to detect the membrane breach even for the maximum percent of cut fibers.

Table 6-2. Sensitivity of various methods based on water quality monitoring (Sethi et al. 2004)

	System					
Methods	**A**	**B**	**C**	**D**	**E**	**F**
Nephelometer	UD	UD	UD	UD	UD	UD
Laser nephelometer	NT	UD	NT	UD	UD	0.002%
Particle counter (≥ 2 μm)	0.001%	UD	UD	0.001%	UD	0.001%
Particle counter (0.5 μm)	NT	NT	NT	NT	NT	0.0001%
Particle counter (0.05 μm)	NT	NT	NT	NT	NT	0.0006%
Particle monitor (filtrate)	UD	UD	0.0003%	UD	NT	UD
Particle monitor (feed/filtrate ratio)	UD	UD	0.001%	UD	NT	UD
Multiple sensing (with laser nephelometer)	NT	NT	0.001%	NT	NT	NT
Maximum percent of cut fibers	0.002%	0.002%	0.0025%	0.001%	0.0025%	0.002%

NT = not tested; UD = unable to detect membrane breach under the condition of the maximum percent of cut fibers.

- Counters that are able to size and count smaller particles (0.05–0.1 μm and >0.5 μm) have higher sensitivity than a standard counter (>2 μm). However, the counter for >0.5 μm showed a better sensitivity than the counter for 0.05–0.1 μm, which might be attributed to the very low counts and small variation under the baseline condition. The counter for 0.05–0.1 μm generated counts two orders of magnitude more than the counter for >0.5 μm. However, the difference in counts for the integral and breached membranes was smaller than that of the counter for >0.5 μm.
- The standard nephelometer was unable to detect the membrane breach for any of membrane systems, even under the condition of maximum breach (0.002–0.0025 percent cut fibers).
- The laser nephelometer and the particle monitor showed intermediate sensitivity. In only one system among the six could the laser nephelometer and the particle monitor indicate a membrane breach.

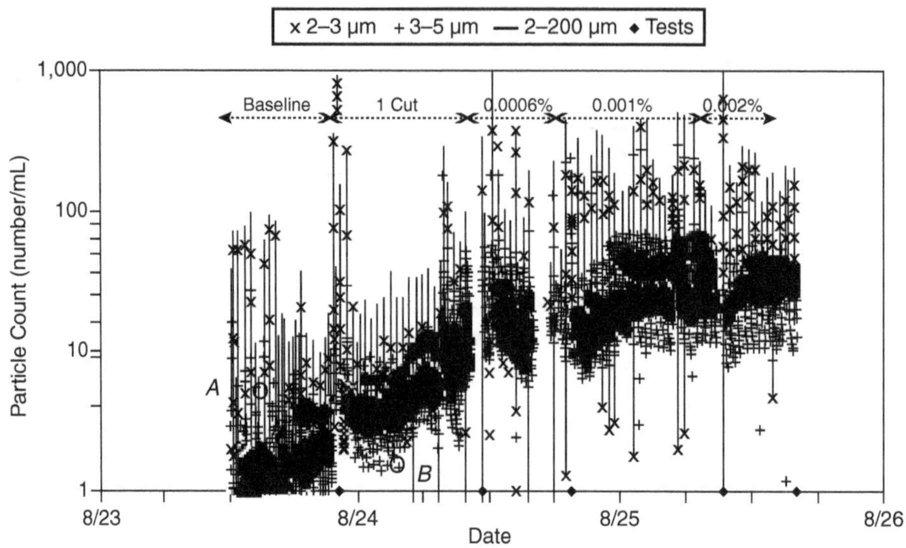

A and B are individual measurements.

Figure 6-22. Particles counts of varying degrees of membrane breaches (Courtesy of Sethi et al. 2004)

Reliability

Another issue is the reliability of the various methods. Reliability is defined as the probability that a measurement truly represents the condition of membrane integrity. Because of the random nature of measurements, two types of errors are likely to occur:

1. A measurement signaling a membrane integrity breach that is actually not a breach (a false positive)
2. A measurement signaling a breach of an actually integral membrane (a false negative)

An example is given in Figure 6-22 in which particle counts (>2 µm) were measured under varying degree of membrane breach. Even though as a whole, particle counts seemed to correlate with the degree of membrane breach, it is difficult to assess individual measurements (such as points A and B in the graph). Therefore, it cannot be determined for certain what it means for each individual measurement, but the probability of whether a measurement signifies an integral or a breached membrane can be assessed.

Because reliability of a membrane integrity-testing method is a measurement of its ability to exclude false signals, reliability can be measured as the probability of false positives or false negatives that are generated by the testing method. The following example illustrates this approach. Assume that *a* and *b* represent the number of positive signals when a membrane is integral and compromised, respectively; and that *c*

Table 6-3. Probability of false positives and negatives of membrane integrity

	Membrane Integrity Status	
Signal	**Integral**	**Compromised**
+	*a*	*b*
–	*c*	*d*
Probability of false positives		$P(\alpha) = \dfrac{a}{a + c}$
Probability of false negatives		$P(\beta) = \dfrac{d}{b + d}$

and *d* represent the negative signals when a membrane is integral and compromised, respectively. These numbers can be used to calculate the probability of false positives/ negatives, as summarized in Table 6-3.

Again, the random nature of the measurement leads to a statistical approach. For a given confidence level, the probability of false positives and negatives can be readily determined if the sample distribution is normal and statistical features of the sample population are known. This may be done essentially by performing a statistical analysis to identify the differences among the sample population (e.g., a *t*-test or an *F*-test), as described in chapter 3. However, because of large overlaps of data points as illustrated in Figures 6-5, 6-11 through 6-13, and 6-15 and 6-16, it is clear that a rigid statistical analysis would result in very large probabilities of false negatives, even though at a relatively low confidence level.

Alternatively, analysis using probability distribution profiles may also be used for such a purpose (Sethi et al. 2004). The approach is based on the thinking that even though the data points have considerable overlaps, as long as the profiles of distribution curves of those measurements can be distinguished from each other, the membrane breaches can still be identified. Using such an approach, the probabilities of false negatives for various methods based on water quality monitoring are illustrated in Figure 6-23. The data for high-sensitivity particle counters (>0.5 μm and 0.05–0.1 μm) and multiple sensors are not included because they were only tested on one membrane system, and there are only a few data points to calculate the probability. This approach may be more robust in the situation of membrane integrity monitoring based on water quality, where the overlaps of data points of integral and breached membranes are considerable and a rigid statistical analysis almost impossible. However, the method's successful application requires the understanding of the impact of water quality change on the profiles of the distribution curves in order to establish the baseline for the measurement.

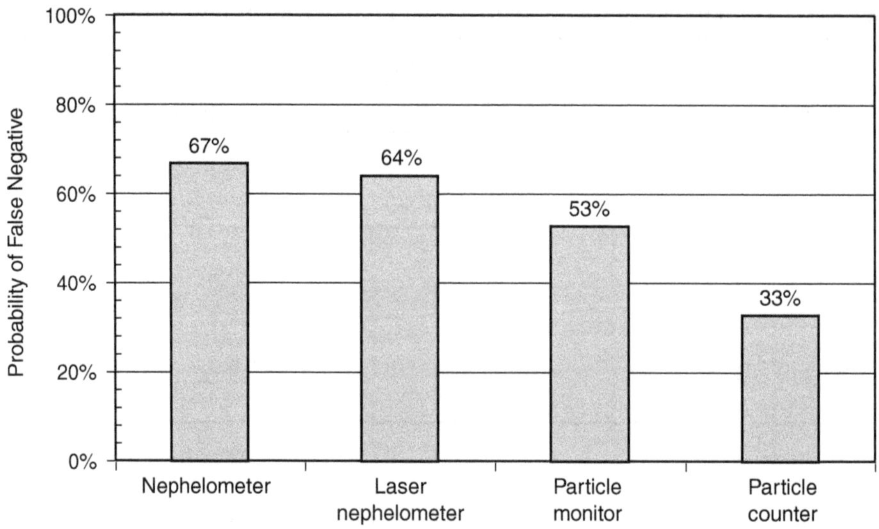

Note: Particle counter data are for particles ≥ 2 μm

Figure 6-23. Probability of false negatives of various methods based on water quality monitoring (plotted based on data of Sethi et al. 2004)

Another indicator for measuring reliability is the monotonic response of a method in relation to the degree of membrane breach (Sethi et al. 2004). By this measure, only particle counters for two membrane systems are able to show monotonic responses. All other methods are unable to do so.

As indicated previously, the reliability of methods based on water quality monitoring is generally poor, as indicated by either the large probabilities of measurements resulting in false negatives or the inability to show monotonic responses in relation to the degree of membrane breaches.

CONCLUSION

The rationale behind requiring membrane integrity monitoring based on water quality is that it can provide continuous monitoring over the period of operation. Therefore, the monitoring methods discussed in this chapter are supplementary to the integrity testing methods based on the flow of fluids discussed in chapter 5. The low sensitivity and reliability of methods based on water quality monitoring would make it difficult to use those methods for integrity monitoring purposes, especially when the membrane breach is small. One exception to these general statements may be the marker-based method, which has higher sensitivity and reliability than other methods based on monitoring water quality but may be considered a continuous method because of practical considerations to maintain. For these reasons, the *Membrane Filtration Guidance Manual* considers the marker-based method as a direct method for membrane integrity testing (USEPA 2005).

These methods have low sensitivity and reliability for two reasons. First, the bypass flow through the membrane breach is highly diluted by the large volume of filtrate (a characteristic of low-pressure membrane systems); and second, because the results of measurements are influenced directly by the quality of the raw water, increasing the sensitivity of detection does not necessarily make measurements more reliable, and it sometimes has the opposite effect. Therefore, improvements to testing methods based on water quality monitoring must address both issues simultaneously. More research is needed in this area—not only developing hardware that generates more sensitive and reliable measurements but also discovering more robust algorithms to interpret the data.

7

OTHER INTEGRITY TESTING AND MONITORING METHODS

INTRODUCTION

In chapters 5 and 6, two categories of integrity testing methods were discussed: those based on fluid flow and those based on water quality monitoring. Those two categories of methods comprise most of the methods used for integrity testing and/or monitoring currently applied to low-pressure membrane systems.

In this chapter, three methods not discussed previously are presented: (1) acoustic method using hydrophones, (2) particle amplifier, and (3) relative trans-membrane monitoring. The acoustic sensor/hydrophone method, which is based on acoustic signals generated during the filtration cycle of low-pressure membrane systems, is completely different from integrity testing methods based on fluid flow and water quality monitoring. Strictly speaking, the particle amplifier and relative trans-membrane monitoring can be classified as water quality based; however, they are unique in comparison to the conventional methods in this category and may be worthy of a place of their own.

The idea common to the methods (2) and (3) is to analyze the accumulated effect—not the instantaneous effect—of membrane integrity breach. This approach may mitigate many pitfalls of conventional integrity monitoring based on water quality and lead to more sensitive and reliable monitoring methods. Although the acoustic sensor has been used to monitor the membrane integrity in a production plant (Petry et al. 2001), the particle amplifier and relative trans-membrane monitoring methods are still in the development stage. This chapter will present the working principles of the methods first, followed by analysis of factors affecting their sensitivity and reliability. Application data will be used for method validation and demonstration.

ACOUSTIC INTEGRITY MONITORING (AIM)

How AIM Works

Acoustic integrity monitoring (AIM) uses an acoustic sensor (hydrophone) to sense and transmit underwater sound signals. A hydrophone works very much like a microphone, its counterpart in the air. AIM technology has its origin in marine voyaging.

There was a written record about use, as early as the later fifteenth century, of an air-filled tube—the predecessor of the hydrophone—to listen to underwater sounds (Urick 1983). Later, hydrophone technology was incorporated into the technology of underwater sonar to detect the approach of enemy submarines. The Allies used this technology successfully to reduce the threat by German U-boats in World War II.

To understand how the hydrophone works, it is necessary to introduce basics of acoustics.

Transmission of Sound

Sound is transmitted in a medium (e.g., water) in the form of a wave that alternatively compresses and decompresses the medium. Therefore, transmission of sound can be described by the change in pressure of the medium over time:

$$P = P^{*}\sin(\omega t) \qquad \text{(Eq 7-1)}$$

Where

P^* = amplitude of sound pressure, Pa

ω = frequency of sound wave, second^{-1}, or Hz

The frequency for audible sound to a human's ear is in the range of 20 to 20,000 Hz. Below 20 Hz, the sound is infrasonic. Above 20,000 Hz, the sound is ultrasonic.

If we observe the pressure over time at one point, it would behave like a curve, as illustrated in Figure 7-1.

Another parameter used for describing sound is wavelength, λ, which is measured in units of length (e.g., meter). The speed of sound, v_S, referred to as the longitude motion of wave front, is related to frequency and wavelength by the following equation:

$$v_S = \omega\lambda \qquad \text{(Eq 7-2)}$$

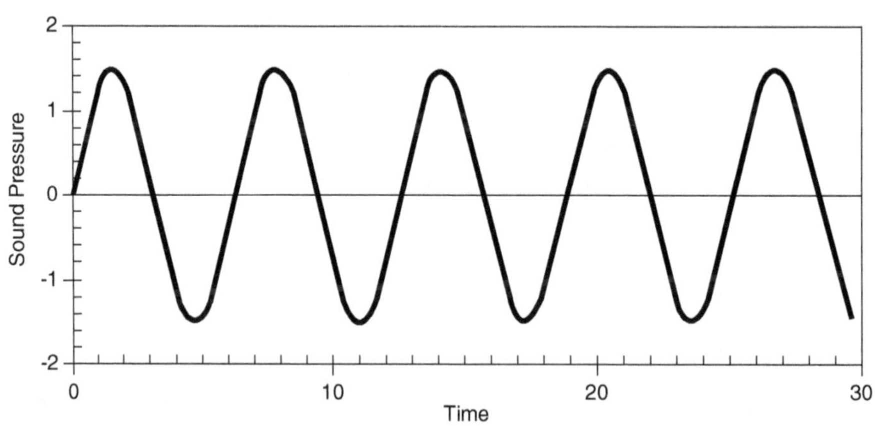

Figure 7-1. The propagation of sound wave

The speed of sound depends on the medium in which it propagates and can be calculated by the following equation:

$$v_S = \sqrt{\frac{C}{\rho}} \quad \text{(Eq 7-3)}$$

Where
C = coefficient of stiffness, kg/m-sec^2
ρ = density of medium, kg/m^3

As the speed of sound is proportional to the inverse of the square root of medium density, sound propagates in water at a much higher speed than it does in air. At 25°C, the speed of sound in freshwater is 1,497 m/s, more than four times that in ambient air. For a single-frequency sound wave, the product of its speed and the density of the medium is called *acoustic impedance.*

Sound intensity, I, expressed as sound energy per unit area of propagation (watt per square meter), can be related to sound pressure by the following:

$$I = \frac{P^2}{\rho \times v_S} \quad \text{(Eq 7-4)}$$

Typically, the level of sound intensity is defined in relative terms—as a logarithm of the ratio of sound intensity to a reference (I_0) of 1 picowatt per square meter (10^{-12} W/m^2). That is,

$$L_I = 10 \log \frac{I}{I_0} \quad \text{(Eq 7-5)}$$

Where
L_I = sound intensity level, dB

Per Equation 7-4,

$$L_I = 10 \log \frac{P^2}{P_0^2} = 20 \log \frac{P}{P_0} \quad \text{(Eq 7-6)}$$

Where
P_0 = reference sound pressure, 2×10^{-5} Pa

Note that sound intensity level is a logarithm: Every increase of 10 dB in sound intensity level (L_I) relates to the increase in sound pressure about three times. To put the numbers into perspective, Table 7-1 lists sound intensity levels of some sound sources and environments.

Very much like light, the strength of the sound signal may be reduced by absorption, reflection, and scattering during the process of sound propagation. Absorption of sound

Table 7-1. Sound intensity levels of some sound sources and environments

Source	Sound Intensity Level (dB)
Hearing threshold of human's ear	0
Rustling leaves	10
Quiet bedroom at night	30
Normal conversation (distance of 1 m)	50–60
Vacuum cleaner (distance of 1 m)	70
Power lawnmower (distance of 1 m)	80–95
Chainsaw (distance of 1 m)	110
Hearing threshold of discomfort for human's ear	120
Jet aircraft (distance of 50 m)	140

NOTE: dB = decibel

depends on the frequency of the sound and the characteristics of the medium. Absorption of low frequency sound is generally low. The major contribution to sound attenuation in water is viscosity. An additional contribution can come from ionic relaxation of certain chemical species in seawater (e.g., $CaSO_4$, $MgCO_3$, and $B(OH)_3$) for sound in a certain frequency spectrum.

Sound may also be absorbed at the fluid–solid boundary as it penetrates the solid phase. Sound reflection and scattering mainly occur at the fluid–solid boundaries. The degree of sound reflection is related to the difference in acoustic impedance between different phases. The larger the impedance, the more sound reflection occurs. Similar to light, sound scattering depends on the relative ratio of the wavelength of the sound wave and the size of the subjects and acoustic reflectivity of the subjects. Equations in a form analogous to the Beer-Lambert Equation for light transmission can be used to describe sound attenuation.

When the sound waves from different sources propagate through the same medium, if the sound pressure is observed at any point of reference, the sound wave can be treated mathematically from the second source as it lags behind the first one. Let P_1 and P_2 represent sound pressures for the first and second sources, respectively.

$$P_1 = P_1^* \sin \omega t, \text{ and } P_2 = P_2^* \sin (\omega t + \delta)$$

Where

P_1 and P_2 = the sound pressure from first and second sources, respectively

P_1^* and P_2^* = the amplitude of the sound wave from first and second sources, respectively

δ = phase lag of sound wave from second source

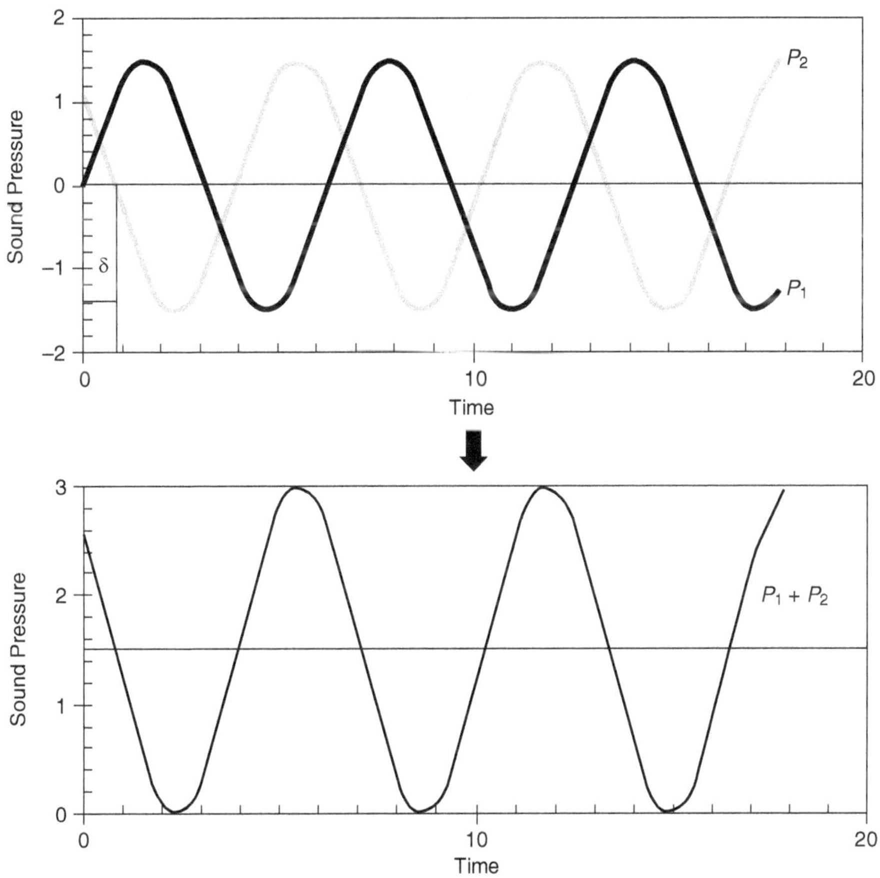

Figure 7-2. Additive nature of sound waves

The sound pressure at the point of reference then is the sum of that from two sources. Figure 7-2 illustrates this concept of the additive nature of sound pressure resulting from multiple sources. For the sake of simplicity, it is assumed in Figure 7-2 that the amplitudes of two sound waves are equal (i.e., $P_1^* = P_2^*$). The phase lag, δ, is critical to determining if two sound sources would strengthen or weaken the combined sound signals. If $\delta = \pi$, the sound waves from two sources with equal amplitude would cancel each other out. If $\delta = \pi/2$, the sound waves from two sources with equal amplitude would yield a signal of constant intensity ($P^2/\rho v_S$).

Hydrophone

Most hydrophones are based on the technology of a piezoelectric transducer, which can generate electrical voltage when subjected to a pressure change (i.e., sound). The transducer can be made with certain crystal, ceramic, or organic polymers. Piezoelectric sensors have been used extensively in ultrasonic transducers for medical imaging and industrial nondestructive testing.

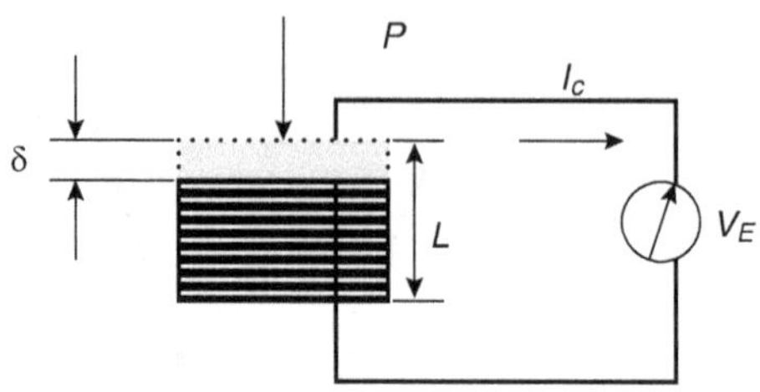

$I_C = k_d P = \varepsilon^P V_E$
$V_E = k_g P$
$k_d = k_g \varepsilon^P$

S = strain, $S = \delta/L$
P = stress (sound pressure)
V_E = electrical field strength (voltage)
I_X = electrical charge density (current)
E = modulus of elasticity
k_d = piezoelectric charge constant of material
k_g = piezoelectric voltage constant of material
e = permittivity (or dielectric constant) of material at constant stress

Figure 7-3. Schematic diagram for working principle of piezoelectric sensor

A piezoelectric material has sites with positive and negative charges that are separated but symmetrically distributed throughout the material. When a mechanical stress is applied, the symmetry in the charge distribution is disrupted, and an electrical field is created across the material. The working principle of a piezoelectric sensor is illustrated in Figure 7-3.

The behavior of piezoelectric materials can be described by coupled equations that relate electrical properties of the material to mechanical stress applied to the material:

$$I_C = k_d P = \varepsilon^P V_E \quad \text{(Eq 7-7)}$$

$$V_E = k_g P \quad \text{(Eq 7-8)}$$

Where

P = stress (sound pressure)
V_E = electrical field strength (voltage)
I_C = electrical charge density (current)
k_d = piezoelectric charge constant of the material
k_g = piezoelectric voltage constant of the material
ε^P = permittivity (or dielectric constant) of the material at the constant stress.

As shown in the equations, k_d, k_g, and ε^P are piezoelectric material properties. High values of k_d and k_g represent high efficiency of conversion from mechanical energy to electrical energy for a material. Consequently, the sensor made of the material has high sensitivity.

Table 7-2. Acoustic integrity testing results (Glucina et al. 1999)

Filtration Mode	Acoustic Signal From Hydrophone: Intact Module (dB)	Acoustic Signal From Hydrophone: Compromised Module With One Cut Fiber (dB)	Difference (dB)
Dead-end	–25	–6	19
Cross-flow	–0.5	29	28.5

Table 7-3. Noise level for common plant equipment (adapted from Corbitt 1990)

Equipment	Sound Intensity Level (dB)
Blowers (forced, induced, fan, etc.)	80–100
Air compressor	95–100
Turbo generators	90
Pumps	80–90
Pneumatic power tools	90–115

Using Hydrophone to Monitor Membrane Integrity

When a membrane breach occurs, it can generate pressure fluctuation due to large pressure differential across the ruptured membrane. This pressure fluctuation becomes the source of sound and can be picked up by hydrophone. Glucina, Lainĕ, Chamant, and Simonie (1999) first implemented a prototype of this technology, named *acoustic integrity monitoring* (AIM), to monitor the integrity of three ultrafiltration plants. Each acoustic sensor was mounted on an individual membrane module, and the signals were sent to a signal measuring device (collector). The collector measured the signal from the sensor array in serial mode and sent the result to a processor to compare it with the baseline. The baseline sound signals of the intact modules were compared with those from a compromised module having one cut fiber. The result, presented in Table 7-2, indicated a significant difference in noise level for intact and compromised modules, although the difference depended on whether the plants were run at dead-end or cross-flow mode. Cross-flow mode generated a more significant difference in noise level (~10 dB). Because the flow through a hollow-fiber lumen is much higher than that at the dead-end mode, a larger pressure fluctuation caused by a ruptured membrane is expected.

The result also showed that the performance of the technology depended on the background noises of the plant. In plants where filtrate flow was regulated by a variable-frequency drive, background noise was lower than in plants that used actuated valves to control filtrate flow. Noises from other sources in the plant also contribute to the performance of this technology by increasing background sound level. Table 7-3 listed the common pieces of equipment in a plant environment and their noise levels.

This technology has been implemented in an 11.4-mgd plant in Lausanne, France. It is reported that the technology is able to detect a membrane breach corresponding to a removal of 6-logs (Petry et al. 2001).

The performance of AIM technology is highly influenced by many system- and site-specific conditions, such as filter configuration (inside-out versus outside-in), mode of operation (dead-end versus cross-flow), and background noise level. Those factors can all influence performance and are difficult to predict beforehand. From a regulatory perspective, it is not clear how the performance criteria such as resolution and sensitivity can be met with this technology. It may be possible to establish an empirical correlation between signal strength and degree of membrane breach via experiments. Multiple sources of noise, either from membrane breaches or from background noise, may be problematic because of the additive nature of sound waves, as previously discussed.

One possible solution is to study the frequencies of sound signals from membrane breaches and install a frequency filter to reduce the interferences from background noises. Because of the empirical nature of this approach, it can be difficult to implement. Another issue is cost. Based on the preliminary estimate for a 28-module rack, the capital cost to implement AIM is approximately $250/module (Glucina et al. 1999)—a very significant amount, although it is comparable to the multiple-sensing technique using laser turbidity measurement.

Overall, the attractive aspect of AIM is its ability to monitor membrane integrity on-line with seemingly high sensitivity. In addition, acoustic sensors on individual modules make it possible to trace on which module membrane breaches occur. More research is needed to understand the relationship between sound level and various factors discussed previously to further improve the usefulness of AIM. In addition, the cost of the hydrophone must be reduced significantly to make the technology more cost-competitive.

PARTICLE AMPLIFIER

Method Description

Because of the low number of particles in the filtrate of low-pressure membrane systems, a direct particle count yields results that are neither sensitive enough nor reliable enough to be used as a monitoring tool for membrane integrity. The particle amplifier addresses this concern. The concept of particle amplifier is straightforward: to place a small backwashable membrane filter downstream of the membranes to concentrate the particles in the filtrate stream. This filter can be backwashed periodically, and the backwash stream can be analyzed by a particle counter. The concept is illustrated in Figure 7-4.

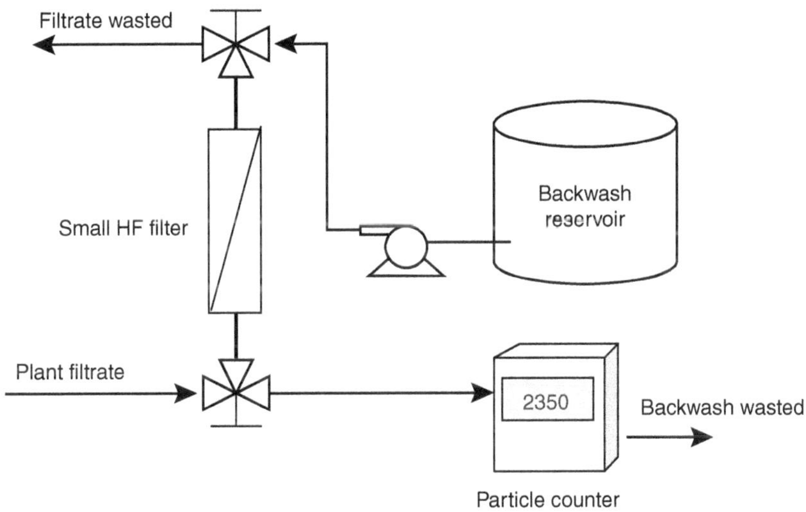

Figure 7-4. Concept schematic for particle amplifier

Because of the concentration effect, the signal strength is "amplified" for achieving enhanced sensitivity. The sensitivity of the particle amplifier can be adjusted by changing the flow rate of filtration and/or the interval of backwash. Because the particle amplifier operates at a batch mode of concentrating-backwash cycle, this method is generally not considered a continuous integrity monitoring method as other methods based on water quality monitoring are.

An important factor affecting the sensitivity and reliability of the method is the recovery efficiency of the small membrane filter. In practice, a small hollow-fiber membrane filter was tested to prove the concept.

Using Particle Amplifier to Monitor Membrane Integrity

The method was tested in two production plants (Sethi et al. 2004). The results are depicted in Figures 7-5 and 7-6. In the box-and-whisker plots, whiskers represent the average particle counts from backwash and the boxes represent 95 percent confidence intervals. For membrane C, particle counts for compromised membranes are higher than the baseline value. However, the counts do not correlate to the degree of membrane breach. For membrane D, results are less clear because counts for compromised membranes are both above and below the baseline value.

Because the membranes used in the particle amplifier have the same or even larger pore size than those of the membranes used in the plants, the particle amplifier should not be able to retain the small particles passing through the intact membranes used in the plants. An obvious question is from where the particles come for baseline readings (intact membranes), as both membranes C and D have membrane pore sizes orders of magnitude lower than 2 μm. It is not clear whether those counts were the artifacts

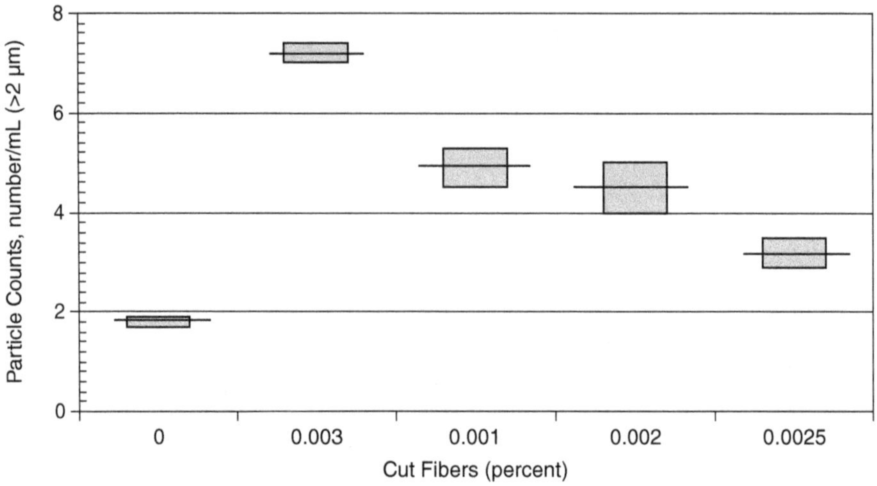

Figure 7-5. Results of particle amplifier for membrane C. The box-and-whisker plot indicates minimum, mean, and maximum, respectively (plotted from data of Sethi et al. 2004).

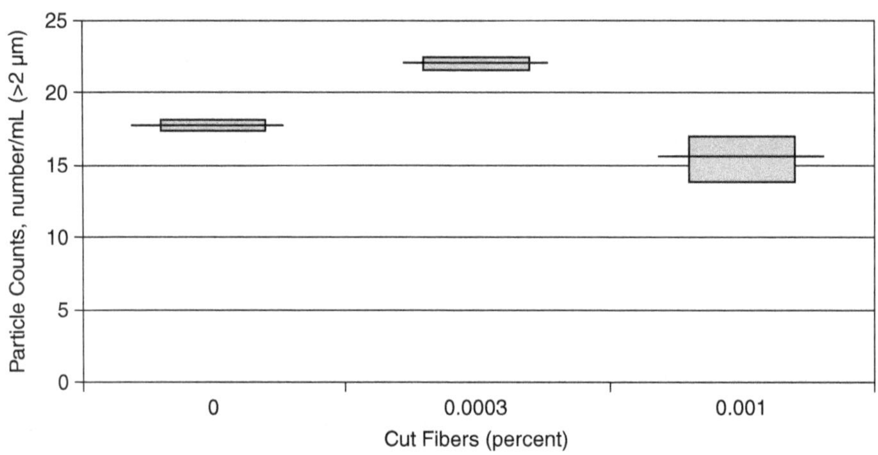

Figure 7-6. Results of particle amplifier for membrane D. The box-and-whisker plot indicates minimum, mean, and maximum, respectively (plotted from data of Sethi et al. 2004).

of the measurements (e.g., particle shedding from backwash), because information about whether a positive control was included in the tests is lacking. Nevertheless, the method does provide a concept potentially applicable to membrane integrity monitoring, although more research is needed to evaluate its feasibility. However, it would be difficult to use this method to provide quantitative information on the degree of the membrane breach in terms of log reduction value because of the dependence of the results on raw water quality.

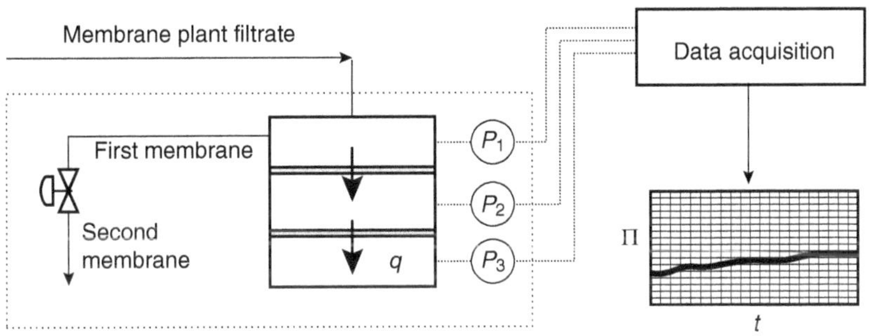

Figure 7-7. Schematic diagram of the device for monitoring relative trans-membrane pressure (TMP)

RELATIVE TRANS-MEMBRANE PRESSURE MONITORING

In one aspect, this method is similar to the particle amplifier method—both measure the cumulative effect of a membrane breach on the filtrate stream, and both involve the use of a secondary membrane device. The difference is that with relative trans-membrane pressure monitoring, the indicator of membrane breach is the increase in trans-membrane pressure (TMP) on the secondary membrane device over time rather than the accumulation of particles in the backwash stream.

How Does the Method Work?

The working principle of the method is to measure the difference in the increase in TMP on the secondary membrane device that is filtering the filtrate stream from a membrane plant (Phattaranawik, Fane, and Wong 2008). A "dirty" filtrate that has been contaminated as a result of the membrane breach should cause more rapid TMP increase of the secondary membrane device than "clean" filtrate, in which membranes remain integral. This method is illustrated in the schematic in Figure 7-7. The device consists of a membrane sensor and a data acquisition unit. The sensor has two membranes arranged in series. Three pressure transducers in the sensor measure the pressures upstream of the first membrane (P_1), downstream of the first membrane (P_2), and downstream of the second membrane (P_3), respectively. Signals from the pressure transducers are sent to the data acquisition unit. There is also a bypass line from upstream of the first membrane of the sensor for adjustment of the feed pressure to the membrane sensor. The method measures relative TMP, as defined below:

$$\Pi = \frac{P_1 - P_2}{P_2 - P_3} \qquad \text{(Eq 7-9)}$$

Where

Π = relative TMP
P_1 = pressure upstream of first membrane of membrane sensor
P_2 = pressure downstream of first membrane of membrane sensor
P_3 = pressure downstream of second membrane of membrane sensor

The purpose for having two membranes in a series is to eliminate fluctuations in flow and pressure in the plant filtrate stream caused by changes in plant operations. As the flow rate fluctuates, the TMP from the first membrane changes. However, as the TMP from the second membrane also changes proportionally to the TMP from the first membrane, the ratio of $(P_1 - P_2)$ to $(P_2 - P_3)$ remains constant.

The method can be used for both constant-flow and constant-pressure modes. At the constant-flow mode, a membrane breach would result in TMP increase caused by particle accumulations on the first membrane. Therefore, relative TMP, Π, would increase over time. At the constant-pressure mode, a membrane breach would reduce the flow to the sensor. Consequently, the TMP from the second membrane would decrease and Π would increase over time. For the purpose of integrity monitoring, either the value of Π or the rate of increase in Π can be used as the monitoring parameter. In addition, composite membranes (membranes with different pore sizes placed in back-to-back fashion) can be used in the places of the first and second membranes in the membrane sensor.

Use of Relative TMP Monitoring for Membrane Integrity

The tests performed using relative TMP monitoring were preliminary. Bentonite was added to the filtrate of a 0.2-μm membrane filter at a concentration of 12.5 mg/L, and the flow rate to the membrane sensor (consisting of two 0.2-μm membranes) was set at 30 mL/min. The response of the membrane sensor (Π value) is depicted in Figure 7-8.

The major variables for the relative TMP method include feedwater quality, degree of membrane breach, flow rate for membrane sensor, and pore sizes of membranes used in the sensor. The higher the particle concentration in the feed, the more sensitive the method and the more rapid the response.

However, it should noted that particle concentration is not the only water quality parameter that affects TMP increase of the membrane sensor. Adsorption of natural organic matter (NOM) on membranes has been identified as the major fouling mechanism for membranes (Lozier et al. 2008). Most NOM has the size orders of magnitude smaller than the pore size of low-pressure membranes used in the industry and can pass through membranes and subsequently adsorbed on the membranes in a membrane sensor.

Therefore, a small membrane breach does not change the NOM concentration in membrane filtrate measurably, but it can affect sensitivity of the method by increasing the

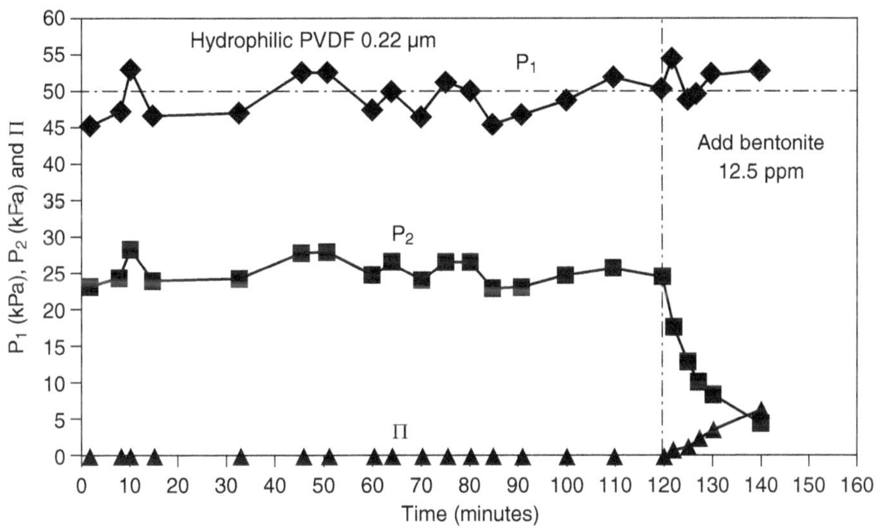

Figure 7-8. Relative TMP method: Result of bentonite spiking test using polyvinylidene fluoride (PVDF) membrane (Phanttaranawik, Fane, and Wong 2008)

baseline value. Moreover, NOM in the membrane filtrate varies seasonally and with the change in pretreatment conditions (e.g., coagulation). Those undoubtedly would increase the difficulty for the method to monitor membrane integrity in a sensitive and reliable manner.

The flow rate to the membrane sensor also can affect both response time and sensitivity. High flow rate would reduce the response time because accumulation of particles on the first membrane of the membrane sensor accelerates. High flow rate can also lower the sensitivity because of high baseline value associated with high flow rate. If the water is indeed free of constituents contributing to TMP increase, the increase in flow rate to the membrane sensor should not affect the baseline value because the ratio of TMP for two membranes in the sensor is constant. However, even for very "clean" flow, the TMP increase over time is still observed and high flow rate would accelerate the increase in TMP. Even though TMP buildup on membranes does not affect sensitivity, it can still need more frequent "reset" for the sensor baseline as the *P* value creeps up more rapidly toward the threshold value over time.

The pore sizes of membranes used in the sensor change the ratio of TMP across the membranes in the sensor. By varying the membranes in the membrane sensor, the sensitivity and response time can be improved. Figure 7-9 compares the membrane combinations of 0.45 µm (the first membrane)/0.2 µm (the second membrane) with composite membranes with a 0.8-µm membrane overlying the first and the second membranes. It indicates for the same bentonite concentration in the filtrate stream, the response time improved significantly in the range of bentonite concentration of

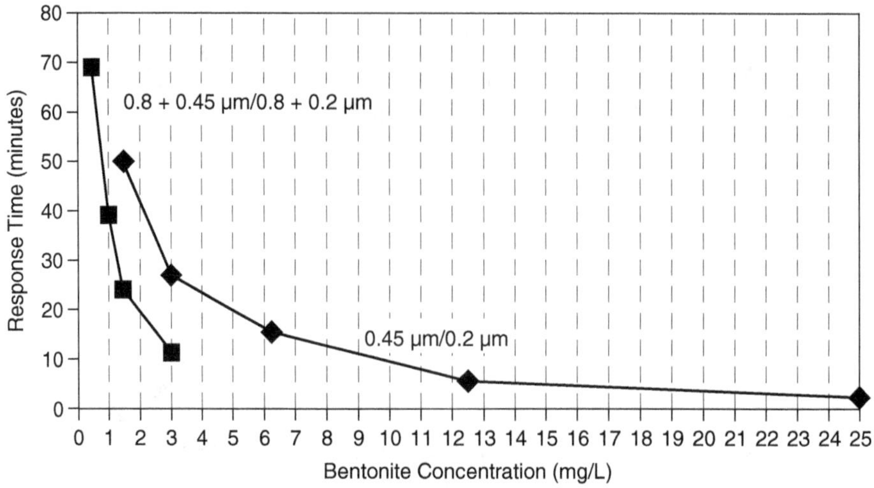

Figure 7-9. Comparison of response time of two membrane combinations as a function of bentonite concentration (Phanttaranawik et al. 2008

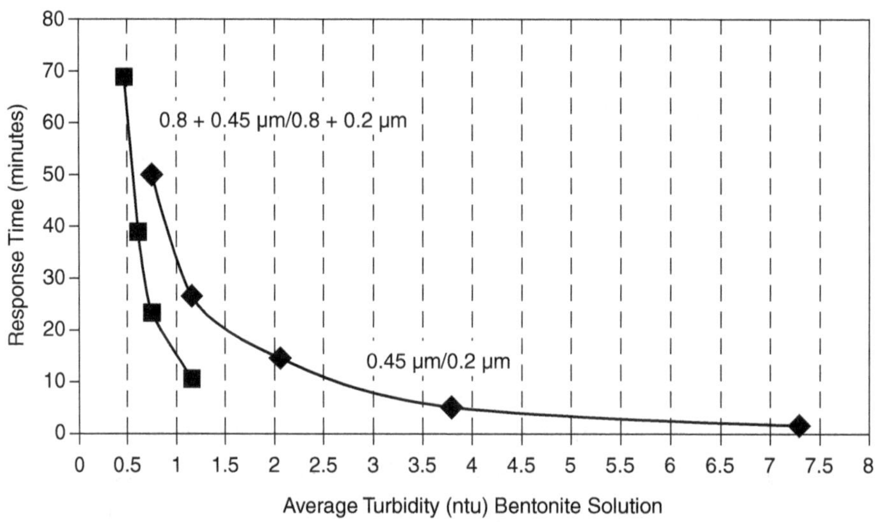

Figure 7-10. Comparison of response time of two membrane combinations as a function of turbidity (Phanttaranawik et al. 2008)

0.5 mg/L to 25 mg/L. Figure 7-10 compares the same membrane combinations as in Figure 7-9, with *x*-axis as turbidity unit instead of bentonite concentration.

The sensitivity of the method still needs to be demonstrated for its application for drinking water. The typical turbidity in membrane plants' filtrate flow is in the range of 0.03 to 0.06 ntu. Even with a membrane breach, the turbidity reading may not change (see Figure 6-5). This may lead to an extended response time according to Figure 7-10.

8

IMPLEMENTING INTEGRITY VERIFICATION AND MONITORING PROGRAM IN LOW-PRESSURE MEMBRANE PLANTS

INTRODUCTION

The ultimate goal for integrity monitoring of low-pressure membranes is to ensure the protection of public health from the risk of drinking water contamination. For the operators of low-pressure membrane plants, implementing a comprehensive program of integrity monitoring and verification is one of the most important operating concerns. According to the US Environmental Protection Agency's *Membrane Filtration Guidance Manual* (MFGM, USEPA 2005), a comprehensive integrity verification program needs to include the following aspects:

- Verifying integrity on an ongoing basis
- Identifying and correcting any integrity problem
- Recording and analyzing integrity test data
- Preparing any required compliance report

For a detailed discussion on what aspects the program should include and how to develop such a program, readers can refer to Appendix A of the MFGM. One example of such a program is discussed from the design perspectives, in which the authors stressed the importance of maintaining an adequate balance of high performance (i.e., log reduction value [LRV]), regulatory compliance regarding the credits granted to the plant, and maintenance requirements in terms of fixing integrity failure (Garcia-Aleman and Lozier 2005).

In this chapter, the discussion is focused on regulatory and technical issues related to implementing integrity verification and monitoring in low-pressure membrane plants. The intention is to help the plant operator understand what regulatory requirements related to low-pressure membranes are and how to comply with them.

DEFINE REGULATORY REQUIREMENTS

The first task for operating a low-pressure membrane plant is to define the regulatory requirements. At the federal level, the most comprehensive requirements for membrane integrity verification and monitoring are reflected in the Long-Term 2 Enhanced Surface Water Treatment Rule (LT2ESWTR) (USEPA 2006a). Several publications have discussed these requirements (Alspach 2007; Liu and Schaefer 2004; Alspach and Allgeier 2003). The following is a brief synopsis of what is required per the LT2ESWTR regarding integrity testing:

- The removal credit for *Cryptosporidium* is granted based on the lower value of (a) product-specific challenge test or (b) system- and site-specific verification via direct integrity.
- To comply with the LT2ESWTR, a membrane plant must undergo periodic direct integrity testing and continuous indirect integrity monitoring (see chapter 4 for the definitions and discussions on direct and indirect integrity testing).
- Direct integrity testing must meet three performance criteria—resolution (<3 μm), sensitivity (site-specific), and testing frequency. A performance-based upper control limit (UCL) must be established to ensure that a system's LRV always meets the granted removal credit for the system.
- Indirect integrity monitoring requires that the filtrate from each membrane unit must be monitored independently and continuously (at ≥15-minute intervals). A performance-based upper control limit (UCL) must be established such that a reading exceeding the UCL triggers direct integrity testing.
- All readings exceeding the UCL must be reported on monthly basis.

The LT2ESWTR gives states certain discretion on regulating membrane systems relating to integrity testing. For example, states can determine testing frequency and approve continuous indirect integrity monitoring technologies they deemed feasible.

At the state level, the approach to regulating membrane systems on integrity issues varies to some degree. For example, WM 880, the guidance document of the Virginia Department of Health (VDH), adheres fundamentally to the regulatory requirements outlined in the LT2ESWTR, but with specific numerical values for sensitivity of direct integrity testing (VDH 2006). However, the California Department of Health Services (CHDS) grants removal credits on a product-specific basis, and requires additional barrier for inactivating and/or removing pathogens and more frequent (direct) integrity testing (Askenaizer, Sakaji, and Jacangelo 2005; Alspach and Sakaji 2007).

In defining regulatory requirements, two aspects are important:

- What is the removal credit that can be granted to a particular low-pressure membrane plant? Is an additional removal credit required?
- What specific requirements for integrity verification and monitoring must be met to be granted removal credit?

With respect to the first aspect, different state regulatory authorities may have different approaches. According to a report funded by the American Water Works

Association Research Foundation (AwwaRF) (Adham et al. 2005), nearly three quarters of 48 low-pressure membrane plants that responded to a survey are granted removal credits 4-logs or lower for *Cryptosporidium*; and nearly 5 percent of 14 plants that responded to the survey are required to have additional log removal ranging from 0.5- to 4-logs for *Cryptosporidium*.

By comparison, nearly 60 percent of 23 ultrafiltration (UF) plants that responded to the survey are granted removal credits of 3-logs or lower for viruses; and over 90 percent of 14 plants that responded to the survey are required to have additional virus inactivation ranging from 2-logs to 4-logs (Adham et al. 2005). Removal credits for *Cryptosporidium* granted by the state regulatory agencies are typically lower than those via either challenge test or integrity testing.

On the one hand, the fact that a large percentage of low-pressure membrane plants are required to have additional removal credits reflects the multiple-barrier treatment concept inherently associated with conventional water treatment processes. On the other hand, granting removal credits for viruses represents a compromise between high removal efficiency demonstrated by challenge tests and the inability of current direct integrity testing to verify removal efficiency. Under the regulatory framework of the LT2ESWTR, granting of removal credits has to be verified via direct integrity testing. Because the size of viruses is much smaller than *Giardia* and *Cryptosporidium*, the testing pressure used in direct integrity testing would be impractical in terms of meeting the resolution criterion. For example, the MS2 phage has a size of approximately 0.025 μm (25 nm). The testing pressure to meet this resolution calculated according to Cantor's Equation (Equation 4-1) needs to be greater than 1,680 psi at 20°C! This is far beyond the pressure ratings for low-pressure membrane systems and impractical to implement. Because verification for virus removal is lacking, state regulatory agencies are generally adapting the multiple-barrier approach, and few states granted virus removal credits to stand-alone UF plants (Adham et al. 2005).

With respect to the specifics for integrity testing and monitoring, available data indicate a diversified picture based on two surveys from two research projects funded by AwwaRF (Sethi et al. 2004; Adham et al. 2005). Figures 8-1 and 8-2 depict the number of plants using different direct integrity testing methods. Figures 8-3 and 8-4 depict the number of plants using indirect integrity monitoring methods.

Although the two surveys were conducted for two separate projects at different times and the surveys' objectives and methodologies were not identical, their results revealed certain common threads:

- For direct integrity testing, results from both surveys indicate that the pressure-hold method is the dominant method used by the majority of low-pressure membrane plants—more than 90 percent of them use it as their primary method for direct integrity testing. Other direct integrity testing methods used

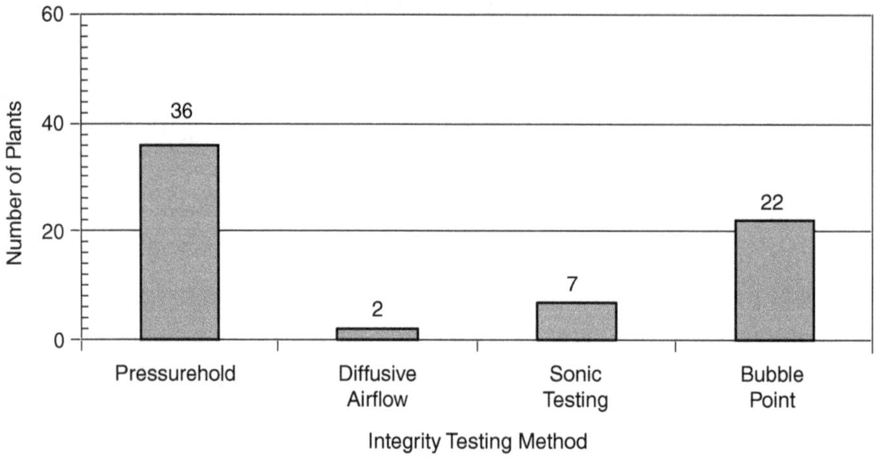

Figure 8-1. Usage of direct integrity testing methods at 39 microfiltration and ultrafiltration plants (adapted from Sethi et al. 2004)

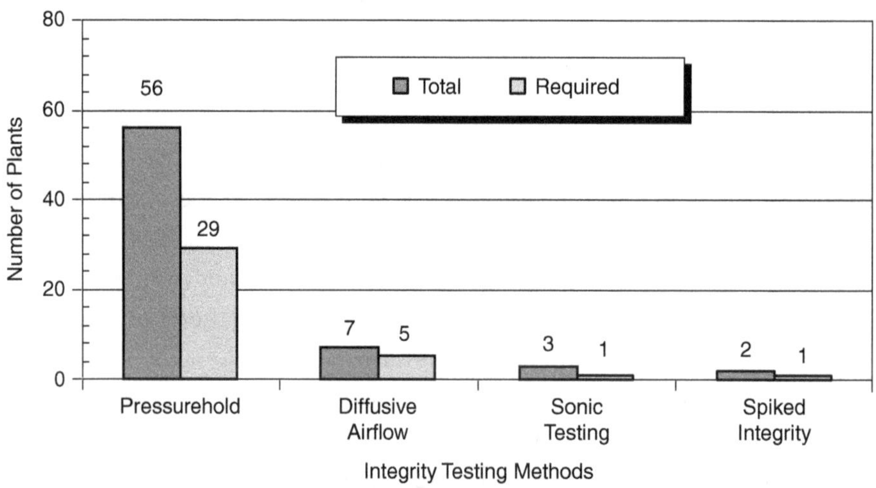

Figure 8-2. Usage of direct integrity testing method in 61 microfiltration and ultrafiltration plants (adapted from Adham et al. 2005)

included diffusive airflow, sonic testing, and spiked integrity monitoring (SIM). With the exception of the diffusive airflow method and SIM method, other direct methods such as sonic testing or bubble-point testing are merely the secondary method for locating membrane integrity breaches.

- For indirect integrity monitoring, although turbidity monitoring is the dominant method (72 percent and 93 percent), particle counts are also widely used (in 44 percent and 67 percent of the plants). Many plants use both methods for integrity monitoring, presumably for both regulatory compliance and enhanced sensitivity of particle counters.

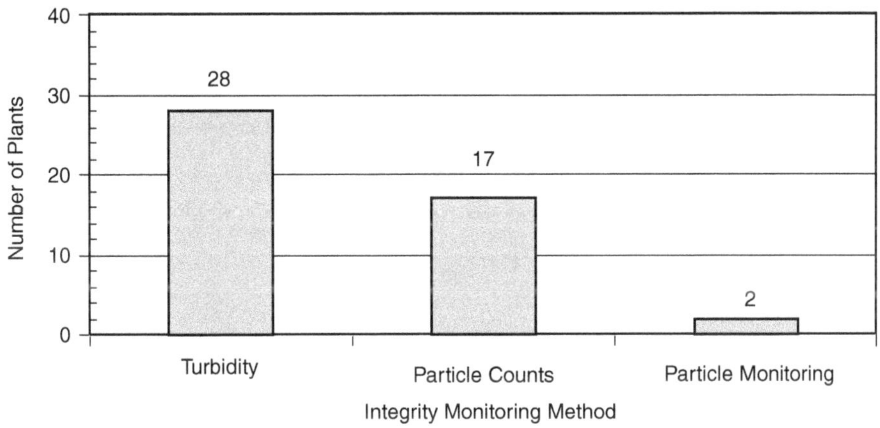

Figure 8-3. Usage of indirect integrity monitoring methods in 39 microfiltration and ultrafiltration plants (adapted from Sethi et al. 2004)

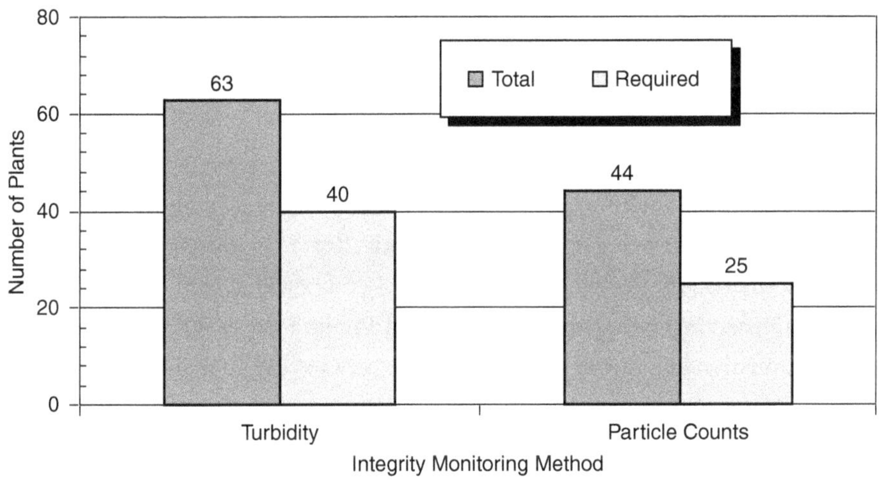

Figure 8-4. Usage of indirect integrity monitoring methods in 68 microfiltration and ultrafiltration plants (adapted from Adham et al. 2005)

As previously discussed, LT2ESWTR does not stipulate specific integrity testing methods, but it requires that the method used must meet certain performance criteria such as resolution, sensitivity, and testing frequency (see chapter 3). In general, meeting the resolution criterion is a relatively simple and less controversial issue and almost all the state regulatory agencies use a similar approach. On the issue of sensitivity, some state regulatory agencies (e.g., Virginia) require a minimum level of sensitivity while others do not. According to the LT2ESWTR, various water sources are assigned to different "bins" based on long-term monitoring results and each bin is required to have certain levels of removal and inactivation of *Cryptosporidium* (USEPA 2006a).

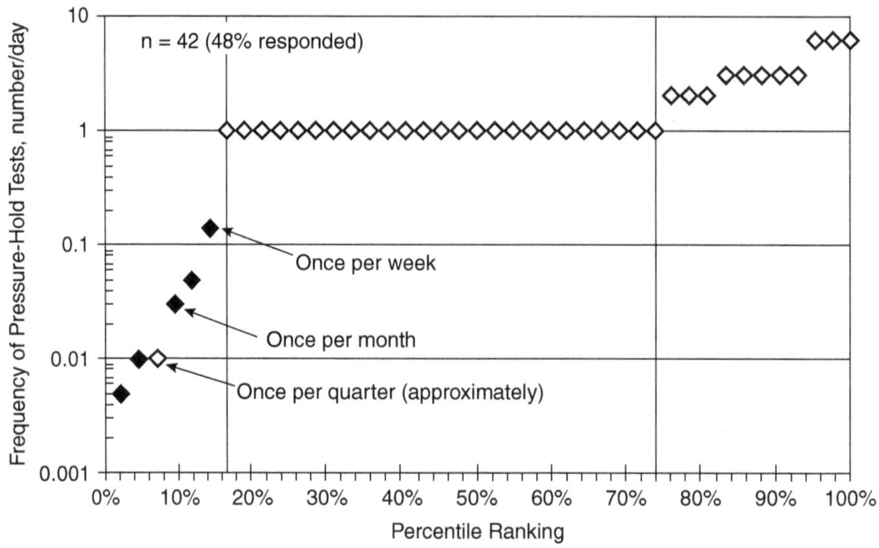

NOTE: Open systems indicate tests required by regulatory agency.

Figure 8-5. Distribution of frequencies of direct integrity testing based on the survey (Adham et al. 2005)

The most controversial issue seems the frequency of direct integrity testing. There are many discrepancies with respect to the frequency of direct integrity testing among the states according to one survey (Adham et al. 2005), as illustrated in Figure 8-5, even though the majority of plants (approximately 60 percent) seemed to test once per day. For example, the California Department of Public Health (CDPH) requires that a newly commissioned plant conduct direct integrity testing four times per day for up to three months and once per day after that (Askenaizer, Sakaji, and Jacangelo 2005).

Alternatively, the Texas Commission on Environmental Quality (TCEQ) requires direct integrity testing once per week and after each chemical cleaning for water sources in Bin 1, and once per day and after each chemical cleaning for water sources in Bin 2 or higher (TCEQ 2008). Such considerable differences in testing frequency reflect different philosophies in balancing the risk of membrane integrity breach against maintaining production of the low-pressure membrane plants because the current practice for direct integrity testing requires the unit to be off-line. From a regulatory perspective, the frequency of direct integrity testing should be based on the risks to the public health rather than the perceived frequency of the occurrence of integrity breaches (USEPA 2006a). An extensive discussion on this subject is included in chapter 3.

It should be noted that both AwwaRF surveys were conducted prior to the publication of the MFGM (USEPA 2005). Since publication of the MFGM, many state regulatory agencies have revised rules on requirements for membrane integrity testing in alignment with the federal regulations. It is important both to consider the current requirements and to anticipate the possible changes in regulations regarding membrane integrity testing. Therefore, defining the regulatory requirements can be a dynamic process that needs to be revisited regularly and improved over time.

ISSUES IN IMPLEMENTING AN INTEGRITY VERIFICATION PLAN

Once the regulatory requirements are defined, operators of low-pressure membrane plants need to establish programs and procedures to comply with those requirements. Among the elements of regulatory compliance are

- Establish procedures and management plans for identifying, recording, and reporting integrity problems.
- Use proper hardware and software to ensure that the integrity testing is in compliance.
- Take corrective actions for noncompliance.

Low-pressure membrane systems provided by vendors typically come with functions to enable both direct and indirect integrity testing. Those functions should be checked out during the plant commissioning, and instruments associated with performing those functions should also be calibrated and tested to ensure that they are in proper working conditions. In addition, the vendor should provide documentation of how the LRV is calculated for direct integrity testing.

A frequently encountered issue is how to determine the baseline value for which membranes are known to be integral and the control limits for direct integrity testing such as the pressure-decay test. Per the discussion in chapter 5, the pressure decay from air diffusion across membranes alone should be rather low as long as one side of the membrane is submerged in water. Therefore, baseline pressure decay where membranes are integral is most likely caused by the minor leaks in the piping system and is not related to membrane integrity. In addition, electrical and magnetic noise in the plant environment can also cause very small fluctuations in readings of pressure transducers.

The baseline pressure-decay rate constitutes the lower practical detection limit for direct integrity testing. In other words, to ensure the reliability of the test, the difference in pressure-decay rate between an integral membrane system and a breached one must be sufficiently significant above the random noise. Every effort should be made to minimize the baseline pressure-decay rate because high baseline value would inevitably decrease the sensitivity and reliability of the test.

The upper control limit (UCL) of direct integrity testing is related to the granted removal credit for a plant. When UCL is exceeded, the membrane unit must be taken off-line for repair. When the sensitivity of a direct integrity test is high, there is sufficient spread between the detection limit and the UCL. A membrane system can be set to have

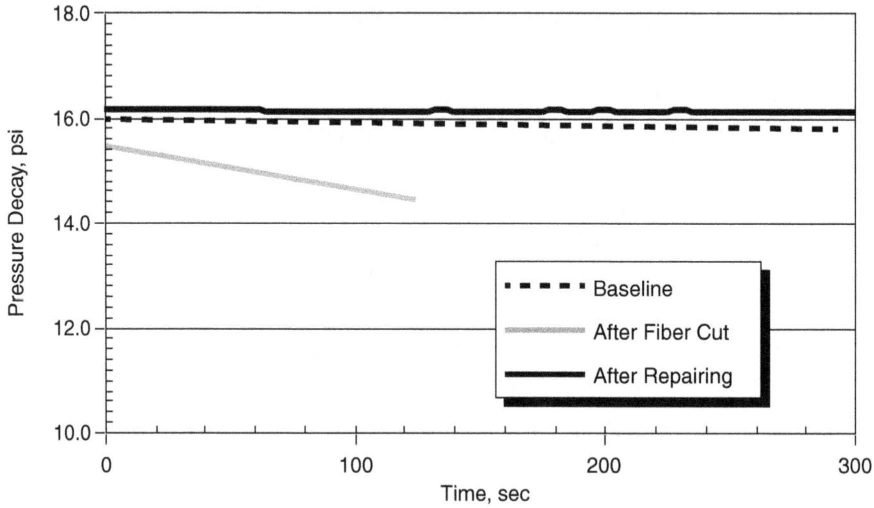

Figure 8-6. Pressure decay of a direct integrity demonstration during pilot test

certain intermediate control limits to alert the operator to a potential integrity breach so that the plant operator can decide if and when the breach is repaired.

Because many utilities conduct pilot tests of the low-pressure membrane process, they sometimes require that demonstration of membrane integrity be included in the pilot program. Such a demonstration typically consists of three parts:

1. Obtain baseline data for membrane integrity testing.
2. Conduct integrity testing on intentionally compromised membranes.
3. Repair the membrane breach in step 2 and conduct integrity testing again to confirm the restoration of membrane integrity.

An example of pressure decay over time in an integrity demonstration program during a pilot test is illustrated in Figure 8-6.

During the demonstration, a single fiber from a hollow-fiber membrane module was cut. The baseline pressure decay over 5 minutes during a pressure-decay test was less than 0.2 psi. After a fiber was cut, the change in pressure decay was rather drastic—approximately 1 psi over 2 minutes, a more than twelvefold increase in the pressure-decay rate. After the breached fiber was repaired by inserting a stainless-steel pin on the filtrate side of the fiber, membrane integrity was restored, as confirmed when the pressure-decay rate resumed its baseline level.

Note that the data from a direct integrity demonstration during a pilot test may not be scalable to a full-scale production plant. This is because the relationship between pressure decay and the LRV depends on many system-specific properties such as flow, hold-up volumes, and operating parameters, which are not likely identical to those used during the pilot test.

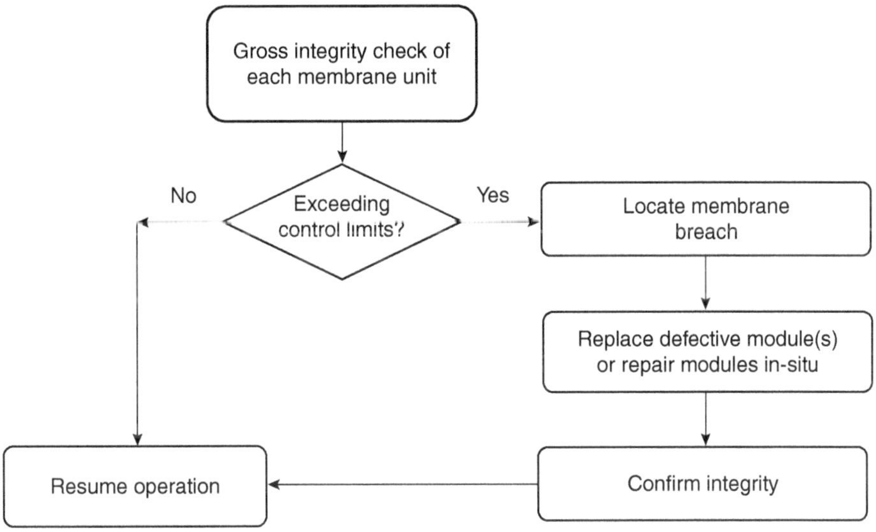

Figure 8-7. Membrane integrity verification system flowchart

Most importantly, a pilot unit is frequently transported to the site and goes through repeated loading and unloading. As a result, the unit is more likely to have air leaks from piping connections. Because the hold-up volume of a pilot unit is considerably smaller than that of a full-scale production unit, the same air leak would have a proportionally more profound impact on the baseline value of the pilot unit. That is, the baseline value of pressure decay for a pilot unit is almost always much higher than that of a production unit.

Therefore, scaling-up from the pilot data is likely to lead to underestimate the sensitivity of direct integrity testing. Ideally, a similar fiber cut test can be performed at the production-plant scale. However, it is often impractical or even impossible to conduct such tests because of constraints posed by local conditions. At a minimum, it is recommended that a plant needs to conduct repeated direct integrity testing in order to fix the air leaks and to establish reliable baseline values during plant commissioning.

Another misunderstanding of LT2ESWTR that is common in the industry is to expect that direct integrity testing of a low-pressure membrane system needs to have the sensitivity to detect a membrane breach of a 3-μm hole. This appears to confuse the resolution criterion required by the LT2ESWTR with sensitivity. In fact, the bypass water flow from a 3-μm hole, even under a worst-case scenario, is so small that it would not have any material impact on LRV.

Procedures and management plans for membrane integrity verification are important components for compliance. Typical procedures include the gross check of the integrity of the individual membrane unit, location of the membrane integrity breach, and corrective actions. A general flowchart for membrane integrity verification procedures is shown in Figure 8-7.

Typically, a gross integrity check is performed on an isolated membrane unit (e.g., a membrane rack). If the result does not exceed the predetermined control limit, the unit resumes normal operation. Otherwise, the system alerts operators to the occurrence of the membrane integrity breach. If the breach is minor and does not exceed the UCL, the system is still in compliance.

However, the operator needs to decide the correct course of action. The MFGM (USEPA 2005) recommends that a membrane unit with detectable integrity breach of any magnitude be immediately taken out of service for a diagnostic test (to locate the breach) and repair. Some state regulatory agencies may also require this. However, if there are no regulatory requirements, the plant operator may chose to continue operation or to take the membrane unit off-line for further inspection to locate the breach and to repair or replace the damaged membrane module(s). The choice depends on the degree of the breach, the need to continue production, and the operating and maintenance schedule.

The procedure for locating membrane breach depends on the specific configuration of a system. For pressurized systems in which each module has a shell, locating the defective module can be done in-situ when there is access to each individual module. Sonic and bubble-point detection are the two methods used most often to locate the defective modules. The sonic method involves attaching an acoustic probe to the shell of a membrane module and listening to the sound generated by air bubbles in the module. Bubble point is a visual method to look for a bubble stream that is a result of a membrane breach. Some low-pressure membrane systems have transparent piping components to allow easy viewing of bubbles. This is shown in Figure 8-8, in which a clear plastic coupling at the top of each module is used to locate defective modules.

Figure 8-9 shows the air bubbles in the plastic coupling caused by a compromised hollow fiber in a full-scale membrane module. They are visible even from a distance of a few dozen feet. For certain configurations of membrane systems (e.g., submerged membranes without a shell for individual membrane modules), locating a defective module may involve disconnecting the system, submerging the suspected multi-module subunit into a separate tank, pressurizing the subunit with air, and looking for air bubbles to locate the exact module with membrane breaches. Easy and quick identification of the defective module(s) can significantly reduce plant downtime.

After the membrane breach is identified, the operator has two options: (a) Replace the defective module and resume the operation, conducting the repair later, or (b) repair the defective module in-situ if possible, confirm its integrity after repairing it, and resume operation. Repairing the defective module is a fairly straightforward process. It typically involves locating the damaged fiber by tracing the air bubbles to the damaged hollow fiber and sealing the damaged fiber(s) by inserting a stainless-steel pin or injecting epoxy resin into the lumen of the damaged fiber.

Figure 8-8. Locating modules with membrane breach

Figure 8-9. Air bubbles from a compromised fiber in a full-scale module

Figure 8-10. Repairing a breached fiber: (a) Locating the breached fiber (by locating where air bubbles come out); (b) pinning; (c) hammering the pin; (d) cutting the pin; (e) endcapping; and (f) reconnecting piping

An example of repairing a breached hollow fiber is illustrated in Figure 8-10. The time required for repair depends on the accessibility to the individual modules in the membrane system. For a membrane system that has easy accesses, the damaged fiber can be located and repaired in the matter of minutes.

It is important to keep track of the history of membrane integrity breaches. The data can be useful for analyzing their causes and helping to formulate corrective actions. Generally, it is plausible to expect that integrity breaches occur more frequently shortly after system startup as the result of either manufacturing defects or improper installation. For example, insufficient flushing of piping may leave construction debris in the system that will damage membranes.

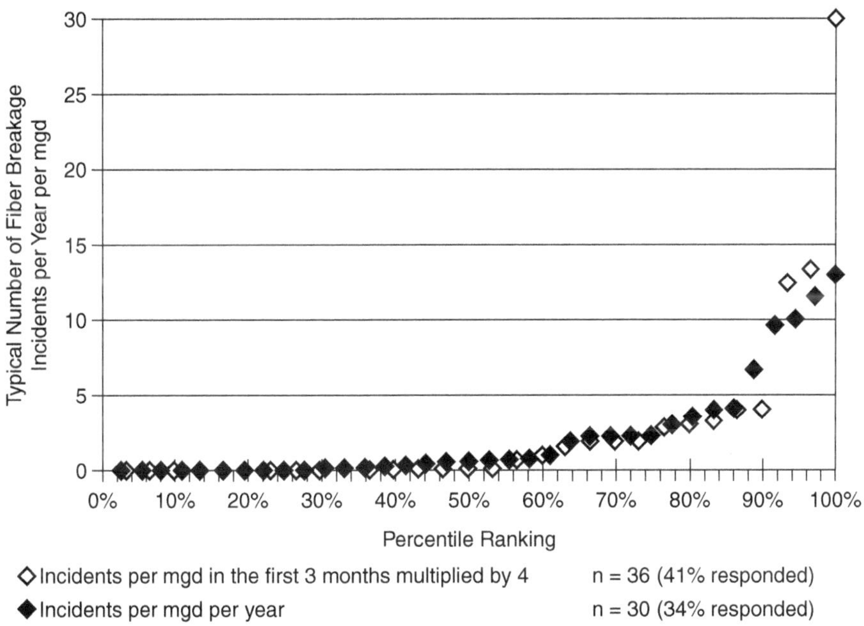

Figure 8-11. Number of fiber breakage incidents reported in an AwwaRF survey (Adham et al. 2005)

However, an AwwaRF survey (Adham et al. 2005) reported that there seemed to be no significant difference in the occurrence of membrane integrity breaches during the first three months of operation and during operation at all the times when data were normalized, as depicted in Figure 8-11. The survey results were based on responses from 36 and 30 plants, respectively—a small portion of low-pressure membrane plants in operation. More data are needed to establish the pattern of occurrence as a function of length in operation.

Another interesting aspect of the survey results is that in the majority of the membrane integrity breach incidents, the number of compromised hollow-fiber membranes per incident appeared to be low. As depicted in Figure 8-12, nearly 90 percent of plants that responded to the survey reported the number of compromised fibers per incident was 10 or less. The low number of compromised fibers per incident is more likely the result of manufacturing defects and/or isolated events (e.g., fibers were cut by a single sharp object in the system).

If the large number of membrane breaches occurs in a short time period, the probability increases significantly that they were caused by factors associated with operating issues. In this case, it is essential to investigate possible causes of breaches to rectify the problems and take steps to prevent it from happening again. Such an investigation is typically performed via module autopsy—looking into the pattern of the membrane breaches, including the spatial distribution of the membrane breaches, the location of

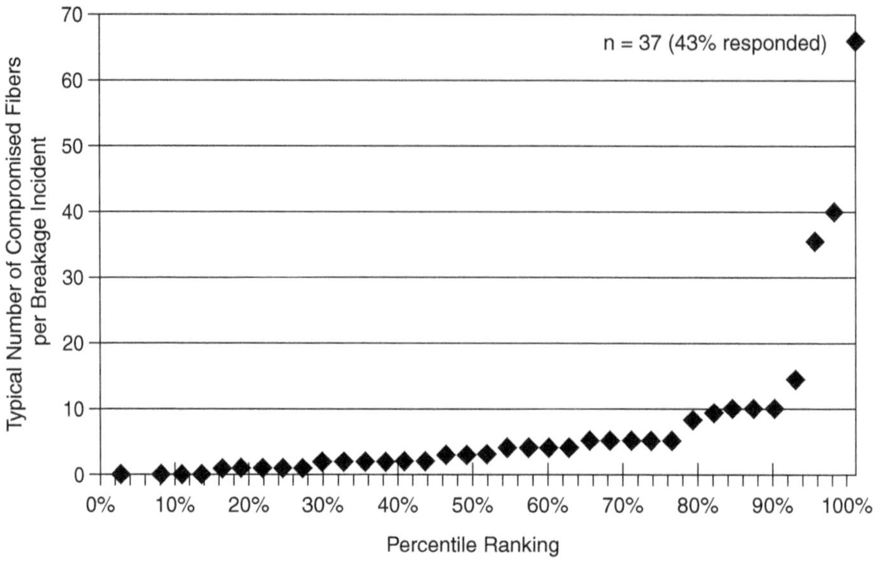

Figure 8-12. Typical number of compromised fibers per incident reported in an AwwaRF survey (Adham et al. 2005)

Figure 8-13. Debris found in membrane module with damaged screen strainer

the breaches in a module, and the morphological features of the breaches. In addition, other physical properties of the membranes, such as tensile strength, elongation, compression strength, and permeability will also be evaluated as a part of the investigation.

The investigation should not be focused solely on membranes. The integrity of the overall system and any operational changes that depart from the normal protocols need to be examined as well. In one incident, it was found that hollow-fiber membranes were damaged by a sharp object that had broken away from a ceramic heating element in the chemical cleaning system. In another example, a damaged screen strainer allowed debris to enter the membrane rack and some of the debris appeared to have sharp edges that damaged the membranes. Figure 8-13 shows the debris collected from an autopsied membrane module.

Appendix 8-A: Resolution and LRV Calculation Using the MFGM Method for Water Treatment Plant

NOTE: Equations and Table are from the USEPA's *Membrane Filtration Guidance Manual,* Chapter 4 and Appendix C. Both are found in Appendix C of this book.

OBJECTIVES

The objective is to determine (1) the testing pressure required to meet the resolution criterion of 3 μm or less as specified in the Long-Term 2 Enhanced Surface Water Treatment Rule (LT2ESWTR), and (2) the pressure-decay value (PDR) corresponding to required log reduction value (LRV) for particles with the size of 3 μm at plant design conditions.

CALCULATION FOR RESOLUTION AND SENSITIVITY OF THE MEMBRANE SYSTEM

1. Determining Testing Pressure for Required Resolution (≤3 μm). The testing pressure can be calculated per Equation 4-1 in MFGM.

$$P_{test} = (0.193 \cdot \kappa \cdot \sigma \cdot \cos\theta) + BP_{max} \qquad \text{(Eq 4-1)}$$

Where

κ = shape correction factor (κ = 1)
σ = surface tension of water (σ = 74.9 dynes/cm at 5°C)
θ = water contact angle of membrane medium (θ = 0)
BP_{max} = the sum of back pressure and static head (BP_{max} = 3 psid)

Substitute the values of parameters above:

$$P_{test} = 14.5 + 3 = 17.5 \text{ psi}$$

Since the testing pressure to be used is above 25 psi and the pressure decay is anticipated lower than 1 psi during the duration of the test for Pall MF system, the resolution criterion is satisfied.

2. Calculating Sensitivity (LRV_{DIT}). The LRV calculation is performed by using Equation 4-9:

$$LRV_{DIT} = \log\left(\frac{Q_p \cdot ALCR \cdot P_{atm}}{\Delta P_{test} \cdot V_{sys} \cdot VCF}\right) \qquad \text{(Eq 4-9)}$$

Where

Q_p = design plant (instantaneous) flow, 918 gpm
VCF = volume concentration factor, 1.08 based on calculated 75th percentile from the challenge test data in AwwaRF Report 91032 (Sethi et al. 2004)

P_{atm} = atmosphere pressure, 14.5 psi
ΔP_{test} = the smallest pressure-decay rate associated with membrane breach, 0.04 psi/min
V_{sys} = the hold-up volume during the test, 47.7 ft^3

The air–liquid conversion ratio (ALCR) is calculated using Darcy Equation by assuming that the hollow fiber breaks completely at the interface of potting layer, which results in a shortest flow path for bypass flow. The calculation also uses the highest trans-membrane pressure (TMP) during a filtration cycle. This results in a conservative result that has a low LRV value.

Air–liquid conversion ratio (ALCR):

$$\text{ALCR} = 170 \cdot Y \cdot \sqrt{\frac{(P_{test} - BP)(P_{test} + P_{atm})}{(460 + T) \cdot \text{TMP}}} \qquad \text{(Eq C-4)}$$

Where:
BP = minimum back pressure plus static head
T = temperature
TMP = maximum allowable TMP
Y = net expansion factor

$$Y \propto \left[\frac{1}{\frac{(P_{test} - BP)}{(P_{test} + P_{atm})}}, K \right] \qquad \text{(Eq C-5)}$$

Where
K = resistant coefficient

$$K = f \cdot \frac{L}{d_{fiber}} \qquad \text{(Eq C-6)}$$

Where:
f = friction factor (f = 0.025)
L = potting length (L = 60 mm)
d_{fiber} = lumen diameter of the fiber (d_{fiber} = 0.64 mm)

The parameters used in the LRV calculation are presented in Table 8-A-1.

Find K:

$$K = 0.025 \times \frac{0.06}{0.00064} = 2.34$$

Table 8-A-1. Parameters used for LRV calculation

Item	Description	Unit	Value
Q_p	Plant design (instantaneous) flow	gpm	918
VCF	Volumetric concentration factor	dimensionless	1.08*
ΔP_{test}	The smallest pressure-decay rate associated with a breach	psi/min	0.04
V_{sys}	System hold-up volume	ft^3	47.7†
BP	Back-pressure during pressure-decay test	psi	0†
T	Temperature	°F	62†
TMP	Terminal trans-membrane pressure during filtration	psi	43.50†
f	Friction factor	dimensionless	0.025
L	The length of flow path for breach	m	0.06
D	Diameter of hollow-fiber lumen	m	0.00064
P_{test}	Testing pressure for pressure-decay test	psi	25.0°

NOTE: *Based on actual microbial challenge data of Sethi et al. (2004)
†Based on the design data

Find Y:

$$\frac{P_{test} - BP}{P_{test} + P_{atm}} = 0.633, \qquad K = 2.34$$

$$Y = 0.625$$

Substitute Y into Equation C-4:

$$\text{ALCR} = 22.2$$

Substitute ALCR into MFGM's Equation 4-9:

$$\text{LRV}_{\text{DIT}} = 4.29 \text{ logs}$$

Therefore, the sensitivity of direct integrity testing is 4.29 logs.

3. Calculate Upper Control Limit (UCL) and Alert Level (AL) for Direct Integrity Testing. The UCL for direct integrity testing, the pressure-decay rate corresponding to the required LRV, is determined by rearranging MFGM's Equation 4-9 yielding MFGM Equation 4-17:

$$\Delta P_{UCL} = \frac{Q_p \cdot \text{ALCR} \cdot P_{atm}}{10^{\text{LRV}^*} \cdot V_{sys} \cdot \text{VCF}} \qquad \text{(Eq 4-17)}$$

Where

ΔP_{UCL} = UCL for pressure-decay rate, psi/min

LRV* = required LRV for the membrane system

The required LRV for the membrane system is 3-logs. Substitute LRV* = 4 and the same parameters in Table 8-A-1:

UCL = 0.78 psi/min

For alert level, use the equation similar to MFGM's Equation 4-17 and substitute LRV* = 3.5,

AL = 0.24 psi/min

The plot of LRV as a function of pressure-decay rate is presented in Figure 8-A-1 in which the UCL and AL are marked with a dotted line.

UCL and AL are indicated on the graph corresponding to LRV of 3- and 3.5-logs, respectively.

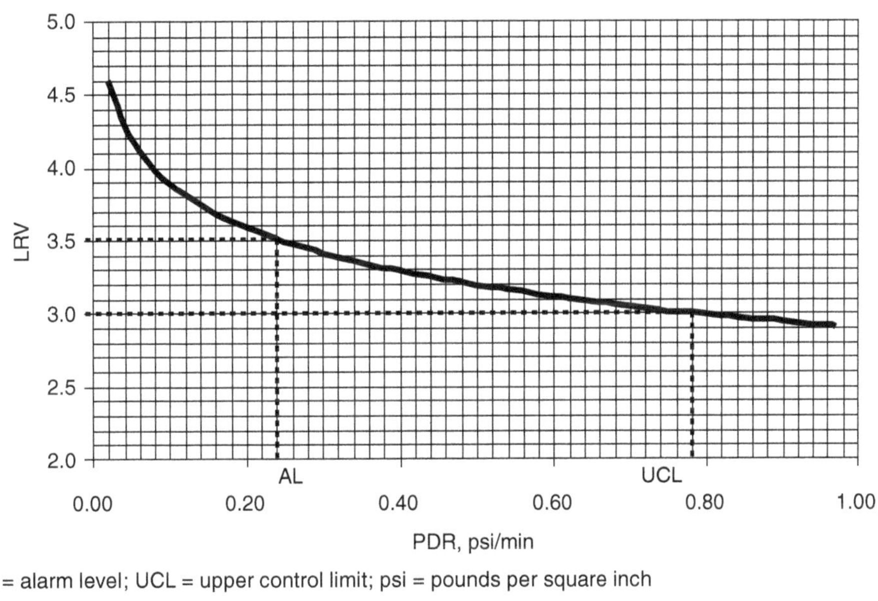

AL = alarm level; UCL = upper control limit; psi = pounds per square inch

Figure 8-A-1 Log reduction value (LRV) as a function of pressure-decay rate (PDR)

9

THE FUTURE OF MEMBRANE INTEGRITY TESTING

INTRODUCTION

The phenomenal growth of the use of low-pressure membranes in the US drinking water industry started less than two decades ago. For an industry that is traditionally conservative and slow in adapting technology, this is almost lightning speed, considering most technologies used in the industry were invented in the 1800s. Therefore, it seems fair to say that membrane integrity testing is not far from its infancy.

The adoption of new technology by the drinking water industry depends on the convergence of regulatory, technological, and economic factors. New regulations for drinking water have been the main driving force for the membrane boom started in the United States from the later 1990s. Of course, regulations are not developed in a vacuum. It does not make sense to have regulations that cannot be complied with because of a lack of available technologies or lack of money. However, technological advances can and often do make new (and more stringent) regulations possible.

In this chapter, the discussion focuses on two categories of technologies that are potentially applicable to membrane integrity testing and monitoring—particle imaging and biosensors. Particle imaging technology is commercially ready for applying in membrane integrity monitoring, whereas biosensor technology is by and large still a work in progress. However, the biosensor technology is very potent and—combined with remote sensing and telecommunication technologies—potentially able to solve many of the problems in the field of membrane integrity testing.

There is no way to tell when particle imaging and biosensor technologies will be adopted, or even if they will. Ultimately, the adoption and deployment of new technologies may reflect the collective willingness of society to allocate resources to achieve its goals—be it delivering safe drinking water, curbing greenhouse gases, or tapping into alternative energy sources.

Regardless of the technologies used, the important issues in integrity testing discussed in chapter 3 are still relevant, although they may be in different contexts. For example, the concept of testing resolution can be interchangeable with specificity if a test can actually identify a contaminant (e.g., *E. coli*) instead of a surrogate (e.g., particle counts).

PARTICLE IMAGING

Particle imaging is analogous to a particle counter but is much more advanced in terms of sophistication and capacity. Instead of pulsing signals generated by light blocking (or scattering) in particle counters, particle imaging uses a microscope and a camera to capture the actual image of the particle. The image is analyzed by powerful image-analyzing software in real time and can be stored for future use. The flow diagram of particle imaging is illustrated in Figure 9-1. The equipment for particle imaging was developed originally for biological and medical research where counting, examining, and sorting microscopic particles suspended in fluid are required.

Human beings are amazingly quick and accurate in recognizing patterns, using eye/brain interaction and our highly developed cognitive process. It is reported that approximately 20 percent of neurons—about 10^{10} of them—in cortex of the human brain are concentrated on vision (Casanova and Ptito 2001).

Duplicating the human cognitive process via computer simulation is extremely difficult and requires a large amount of computing power. Imaging analysis, first developed for the military for use in spy satellite images, is a subset of the study of artificial intelligence and pattern recognition. Even the current best pattern recognition program appears to be primitive and slow compared with a human's brain/eye combination. However, for the relatively simple task of particle imaging, the technology is commercially available.

In particle imaging, the "raw" images are converted digitally into two types of images: binary images and gray-scale images. Binary images contract the areas occupied by a particle and its surroundings. The working principle of binary image conversion is illustrated in Figure 9-2, in which the value of each pixel on the image is compared with a threshold value. If the value is greater than threshold value, it is deemed as particle. Otherwise, it is considered background. Once a raw image is converted into a binary image, geometric attributes such as size, area, length, width, and perimeter of a particle can be measured.

Because most particles have irregular shapes, the parameter representing the size must be normalized. Equivalent spherical diameter (ESD), the most used parameter in this regard, permits the use of a single number to quantify the size of particles of any shape. In addition, there are several secondary measurements to characterize a particle, such as aspect ratio (the ratio of the maximum dimension to the minimum dimension of a particle), circularity, elongation, convex perimeter, and roughness.

The gray-scale image represents 256 discrete levels of intensity (each level is an 8-bit number) and gives a particle optical attributes such as transparency and average

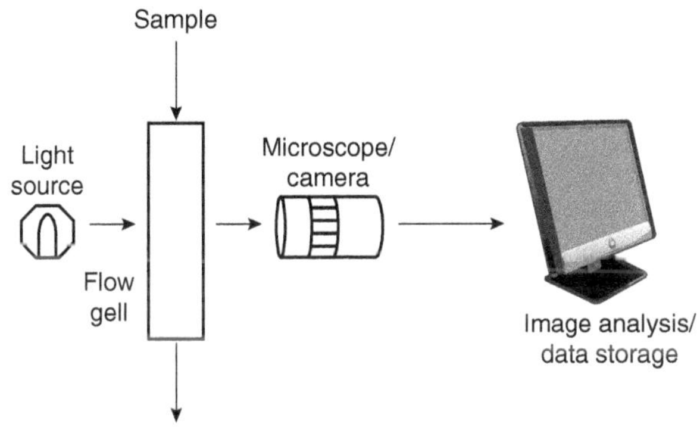

Figure 9-1. Principles of particle imaging technology

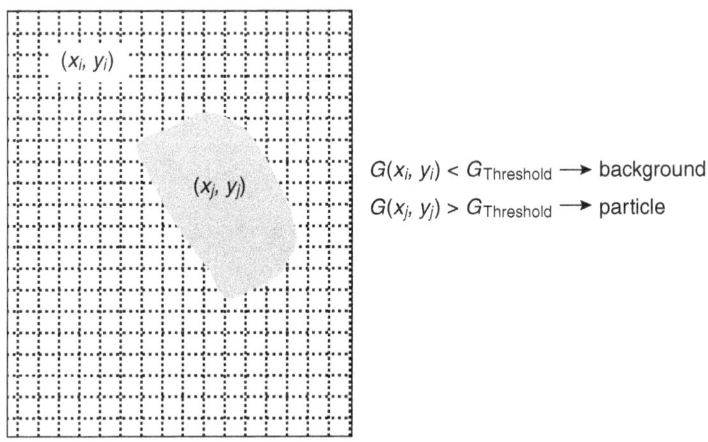

Figure 9-2. Digitization of particle image into binary image

intensity. If the image is colored, a color scale can be similarly constructed based on the three primary colors. Once the image of a particle is captured and analyzed, it can be identified and categorized by comparing its attributes to preselected criteria through data filtering. Those data filters can be either binary (belongs versus does not belong) or probable (probably belongs).

When particle imaging is employed on a continuous basis, especially on a clean fluid, the scarcity of particles may mean that many image frames are empty of particles. Those images are useless as they lack information about particles. To prevent those particle-free images from being analyzed and stored, certain ancillary optical sensors can be used to "trigger" image analysis and storage. That is, an image frame is taken by a camera only when the ancillary sensors detect the presence of particles. This technique is termed frame-on-demand particle imaging. The schematic diagram of such an arrangement is depicted in Figure 9-3.

Frame-on-demand particle imaging greatly reduces the number of images to be analyzed and stored. This saves on equipment cost and, most importantly, makes it

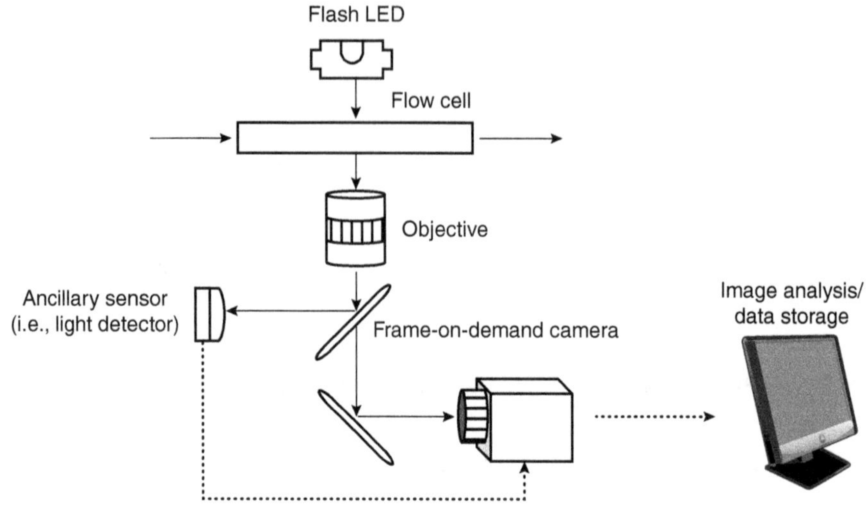

Note: LED = light-emitting diode

Figure 9-3. Frame-on-demand particle imaging

possible to employ arrays of such ancillary sensors to target particular applications (e.g., polarized light, phase contract, and fluorescence sensor).

In addition, the image analysis software is essentially open-structured and can be trained on an on-going basis to recognize and sort particles according to geometric and optical attributes. This opens up huge opportunities for the applications of the technology in many different fields.

Particle imaging combines the advantages of traditional particles counts (speed, simplicity, and real-time analyses without need for sample preparation) and microscopic analysis (accurate sizing, ability of sorting particles, and broad size and concentration range, independent of particle size and optical characteristics) (Thomas and Moore 2003). This enhanced technology is able not only to provide quantitative information such as counts and sizes of particles with greatly improved accuracy but also to generate qualitative information about shape and optical characteristics that leads to identification of the origins of the particles. Therefore, it is a natural tool for diagnosis and optimization of water treatment plant operations (Booth, Carlson, and Kawamura 2006; Philipi et al. 2005; Douglas et al. 2004; Thomas and Moor 2004; Rabinski and Thomas 2003) and for detection and enumeration of waterborne pathogens (McCuin and Clancy 2005; Thomas 2005).

In one application, particle imaging was used to detect a potential algal bloom at an early stage to minimize taste and odor complaints (Reilley-Matthews 2007). The image capture and analysis were triggered by fluorescence signals, and the image analysis software was trained to recognize the species of algae that cause taste and odor complaints by comparing the images of particles with those in the built-in library. The early detection of problematic algal species leads to more efficient and timely algal control

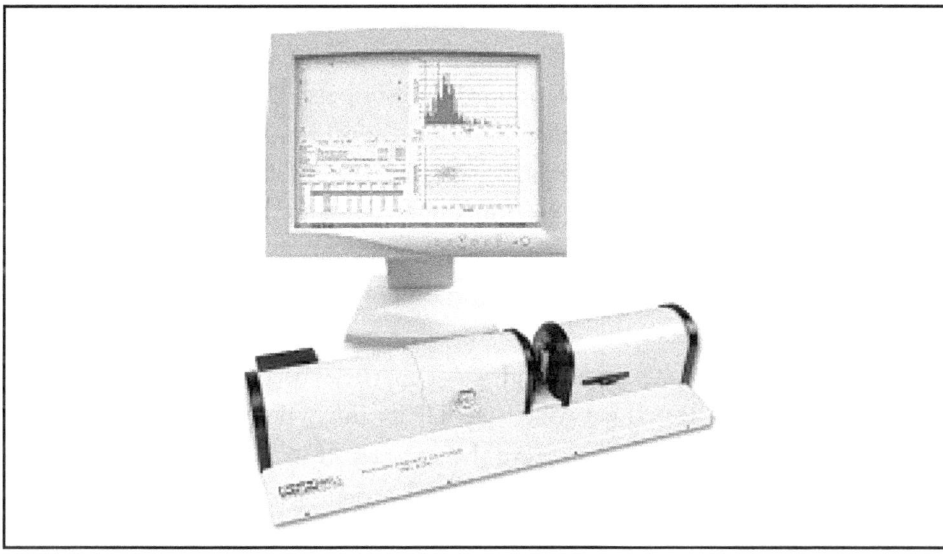

Figure 9-4. Micro-Flow imaging system (courtesy of BrightWell Technologies, Inc.)

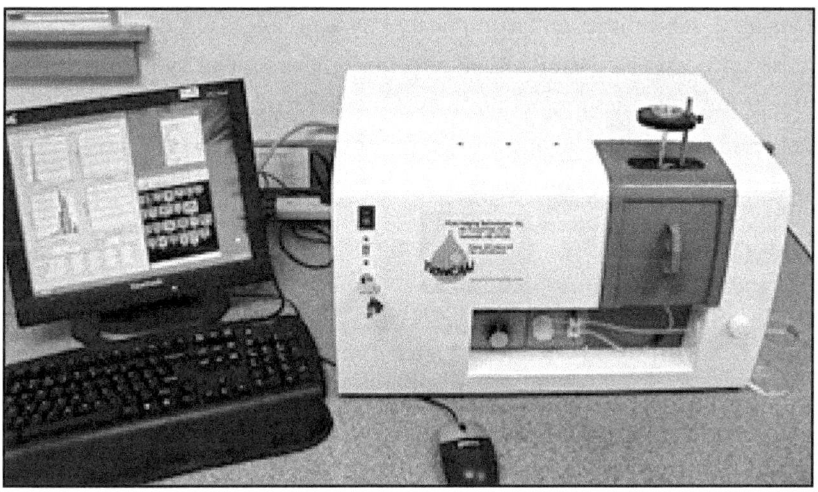

Figure 9-5. FlowCAM® system for particle or cell measurement in solution (courtesy of Fluid Imaging Technologies)

strategies that prevent algal blooms. Consequently, the application of the technology minimized the taste and odor complaints caused by the decay of those algal species. Two commercially available systems for particle imaging are depicted in Figures 9-4 and 9-5, respectively.

In the context of membrane integrity monitoring, the enhanced capacity to categorize particles via particle imaging is very significant. It is conceivable that particle imaging can distinguish particles from microbubbles in the filtrate. As a result, the true count of particles can be generated and used as the indicator of membrane breach. The sensitivity and reliability of membrane integrity monitoring can be improved further if the image analysis software can be trained to distinguish indigenous particles from the feedwater and those from the other sources (e.g., shedding from valves and piping).

The enhanced sensitivity and reliability depend on the ability of the particle imaging technology to distinguish particles from different origins, i.e., the specificity of the measurement. Such specificity is possible only if the range of values of attributes of a particular category of particles and how they differ from other categories of particles are understood. In general, the more attributes that are included in categorizing a particle, the more specific a particular category of particles becomes. Software architecture using an artificial neural network is feasible for this type of work, as it continues to "learn" from the process and improve the accuracy of its projection. Even though groundwork is necessary to define the criteria for various attributes for particles from various origins, there is no inherent technological barrier for applying the technology to membrane integrity monitoring.

A limitation of the particle imaging technology could be traceability, i.e., the ability to identify which module has an integrity breach because the measurement is performed using a centralized and complicated system. As a result, this technology is likely to be confined to continuous monitoring of gross integrity of a membrane unit consisting of multiple membrane modules. Once a membrane breach is confirmed by the use of this technology, the operator can apply secondary detection techniques (e.g., bubble point or sonic detection) to trace the breach to individual module(s).

BIOSENSOR

A biosensor is a device that detects an analyte by combining a biological recognition component with a (physicochemical) signal transducer, using associated electronics to process signals and display the results. The biological recognition component is the sensing unit for the presence of the analyte; it can be of many forms, for example, nucleic acids, antibodies, enzymes, cell receptors, organelles, microorganisms, or even tissues.

The detector (or transducer) component transforms the signal from the sensing unit to another signal that can be easily measured and quantified. Transducers can be electrochemical, optical, or acoustic (piezoelectric).

Biological Recognition

Biosensors are based on three types of recognition mechanisms: (1) bio-catalysis, or enzyme-catalyzed biochemical reaction, (2) bio-affinity, and (3) microorganism-based biochemical reactions. The major types of biological components (and associated mechanisms for analyte recognition), adapted from Rogers and Mascini (2007), are summarized in Table 9-1.

Table 9-1. Biosensors for environmental applications—biological recognition mechanisms (adapted from Rogers and Mascini 2007)

Biological Recognition Component	Reaction
Bio-catalytic (enzyme reactions)	
• Catalytic transformation of analyte into a sensor-detectable form	Enzyme + analyte → product
• Inhibition of enzyme activity by an analyte	Enzyme + analyte – X → product
Bio-affinity	
• Antibodies: compounds or class-specific affinity toward the analyte	Antibody + tracer + analyte → antibody – analyte complex + tracer/pollutant complex
• Nucleic acid: detection of analyte-induced DNA damage or detection of microorganism by hybridization to species-specific sequence	
Microorganism-based	
• Analyte-induced cellular respiration	Analyte (substrate) + test organism → sensor-detectable response
• Inhibition of cellular respiration by analyte	Substrate + test organism + analyte – X → sensor-detectable response
• Promoter recognition by specific analyte followed by gene expression, enzyme synthesis, and catalytic activity	Analyte + test organism → sensor-detectable response

Bio-catalysis

Bio-catalysis is an enzyme-catalyzed biochemical reaction that can be based on either the catalytic transformation of an analyte or the inhibition of enzyme activity by an analyte. In the catalytic transformation mode, the analyte acts as substrate for the enzyme. Because the enzyme/substrate interaction is usually very specific, biosensors based on this recognition mechanism have a very high specificity of detection and are not affected by environmental interferences.

This type of biosensor can be configured to operate continuously and reversibly, an obvious advantage for a system that needs long-term continuous monitoring. The disadvantage of this type of biosensor is its inherently relatively high detection limit, which may be in the micromolar (µM) range. It is possible to improve the detection sensitivity. For example, catalytic cycling of enzymes among various oxidation states can sometimes significantly amplify the sensor response.

Biological recognition based on the enzyme inhibition mode uses the analyte to inhibit the activity of an enzyme. Because inhibition of enzyme activity may involve a large number of analytes, usually of a particular chemical class, the detection is more broadly ranged and less specific. The detection limit based on enzyme inhibition, especially from irreversible inhibitors, can be as low as micrograms per liter (μg/L) and even nanograms per liter (ng/L). Therefore, the sensitivity of the measurement is generally very good. The limitations of biosensors based on this mechanism include the following:

- There is a need for substrates and sometimes for cofactors and mediators to maintain the enzyme as active.
- The analyte may have to be chemically oxidized to metabolic intermediates to gain the maximum sensitivity.
- The irreversible nature of analyte/enzyme interactions means the sensor can only be used once and must be replaced.

Those limitations may render the biosensors based on enzyme inhibition less feasible for long-term monitoring at a remote site.

Bio-affinity

Biological recognition based on bio-affinity has two levels: immunochemical (antibody/antigen) affinity and nucleic acid–based affinity. Immunochemical affinity is more widely used because of the availability of many antibodies that have affinity with a wide range of analytes. Several assay formats can be employed for detection of various analytes. For example, many low molecular weight organic molecules are usually detected by using competitive binding of the analyte with a tracer-antigen on an antibody binding site. This is because those molecules have few distinguishing optical or electrochemical characteristics themselves, so a tracer is needed to signal the presence of those analytes. The tracer can be labeled on either the analyte or the antigen. Either way, the tracer induces an optical or chemo-optical signal that can be detected. This format is often used for fluorescence-based detection systems.

When the antibody is the same as analyte, the system can be simplified. In this case the tracer may no longer be needed, because an antibody is typically a large molecule and its binding to the surface can be detected by the various signal transducers. This format can also be employed in the opposite manner by immobilizing antigens on the surface of the signal transducer. Analytes compete with immobilized antigens for the antibodies. As a result of competitive binding, analytes reduce the antibody/antigen binding, which in turn generates signals that can be detected by various signal transducers.

Another commonly used assay format involves the use of an enzyme-labeled antigen-tracer competing with analytes for immobilized antibody binding sites. The enzyme catalytically converts an undetectable substrate into an electrochemically or optically detectable product. The competitive binding of analytes on the antibody reduces the binding of the enzyme on the antibody and consequently the amount of detectable products. Due to the unique binding between antibody and antigen, the biosensor based

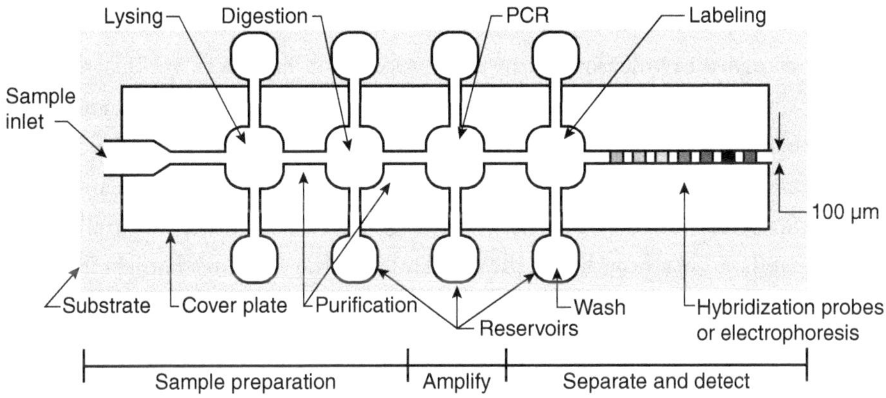

Figure 9-6. Conceptual design of "lab on a chip" for DNA analysis (Jacobson 2007)

on immunochemical recognition can be rather specific. Immunochemical sensors can be made in either disposable or multi-use format. The latter is especially useful for long-term continuous sensing in remote locations.

Another bio-recognition mechanism based on bio-affinity is the use of nucleic acid probes, mostly DNA probes. Because nucleic acids contain genetic information in a sequence of four bases, they can be used to identify organisms by the signature segment specific to each species of organisms. The process that "reads" the order of bases contained in a nucleic acid string is called sequencing. The traditional sequencing process involves multiple steps including sample preparation, amplification via polymerase chain reaction (PCR), separation, labeling, and detection via gel electrophoresis. This process would not be suitable for the task of real-time monitoring due to its complexity and time-consuming nature. However, a recent development in DNA analysis is the concept of "lab on a chip" (Figure 9-6), which would shorten the time needed for the analysis from hours to minutes.

A more promising technology of DNA probes for the application is sequencing by hybridization. The technology is based on the fact that two strings of DNA segments are bonded together by the complementary base pairs they contain—i.e., the process of hybridization. Because the purpose here is to detect the presence of organisms, determining the exact base sequence of DNA is not required. If hybridization occurs with a DNA signature segment of a known organism and the hybridization process can produce a detectable signal, we would know that the organism is present in the sample. Labeling the DNA segment with optical or photochemical tags can be a way to enable detection. Another possibility may involve coating the segment onto a microcantilever. The technology still needs significant developments before it is ready. However, the preliminary results look promising (Cheng et al. 1998).

Microorganism-Based Biosensors

The broad category of microorganism-based biosensors embraces various recognition mechanisms. In one example, analytes may act as substrate for the microorganism to induce cellular respiration, and conventional measurement such as biological oxygen demand (BOD) can be used for detection. Another mechanism involves the inhibition of cellular respiration by an analyte. These two recognition mechanisms are parallel to those of enzyme-catalyzed reactions but at the cellular level. The third mechanism involves the use of promoter recognition. A promoter is a region of DNA that facilitates and regulates the transcription of a particular gene. The microorganisms used in the biosensor are typically genetically modified to contain a promoter that responds to a specific analyte by inducing the gene expression, followed by enzyme synthesis and catalytic activities.

Signal Transducers

A signal transducer is a device that converts one type of signal to another. For example, a pressure transducer converts force (pressure) into an electrical signal. Classified by type, transducers are electrochemical, optical-electrical, or acoustic. Each type of transducer may transmit different types of signal. The major types of transducer used in biosensors for environmental applications, adapted from Rogers and Mascini (2007), are summarized in Table 9-2.

In theory, any biological recognition mechanism can couple with any signal transducer, given an appropriate operating format. However, certain combinations may have inherent advantages (Rogers and Mascini 2007).

Electrochemical Transducers

Electrochemical transducers can be divided into subtypes of signals: potentiometric, amperometric, and conductance. A potentiometric transducer relies on the relationship of electrical potential and the concentration of a charged analyte as described in the Nernst Equation. Potentiometric measurement requires a reference electrode, a special electrode designed to measure the specific ion of interest, and device to measure electrical potential. The potentiometric transducer is used widely in chemical sensors such as pH or ion-selective electrodes. As an example, the chloride concentration of a solution can be measured using a cell assembly as depicted in Figure 9-7, in which chloride ions in the solution react with Ag on the Ag electrode to form silver chloride and electrons flow from the Ag electrode to the reference electrode. The electrical potential across the pair of the electrodes is related to the chloride concentration in solution phase by the Nernst Equation (Figure 9-7). An electrode specific to a charged species can be used as the signal transducer. The main advantages of such device include the wide concentration range (generally on the order of 10^{-6} to 10^{-1} mol/L), ability for continuous monitoring, and low cost. However, electrodes may not be very selective, may be prone to environmental interference, and may have relatively high detection limits (Rogers and Mascini 2007).

Table 9-2. Biosensors for environmental applications—transducers (adapted from Rogers and Mascini 2007)

Transducer	Type of Signal
Electrochemical	Potentiometric
	Amperometric
	Conductance
Optical/electrical	Light addressable
	Potentiometric
	Surface plasmon resonance
	Absorbance
	Luminescence
	Fluorescence
	Total internal reflectance fluorescence
Acoustic	Quartz crystal microbalance
	Surface acoustic wave
	Surface transverse wave

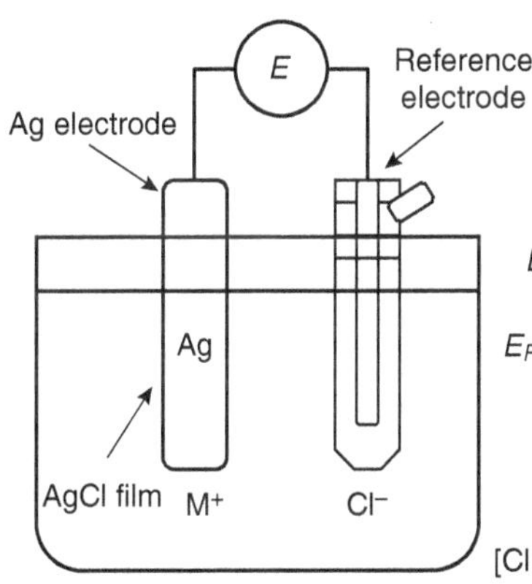

$$AgCl + e^- \rightarrow Ag + Cl^-$$

$$E = E^0 - E_{Ref} + \frac{RT}{zF} \ln [Cl^-]$$

E = electrical potential across electrodes
E^0 = standard electrical potential for chloride electrode
E_{Ref} = electrical potential for reference electrode
R = universal gas constant
T = absolute temperature
z = charge of the ions
F = Farady constant
$[Cl^-]$ = chloride concentration in solution

Figure 9-7. Potentiometric analysis for chloride

Many enzyme-catalyzed reactions are reduction–oxidation (redox) reactions that involve the transfer of electrons from the analyte to the product. When the electrons flow from the special electrode to a reference electrode, the current is proportional to the concentration of the product of the redox reaction, which in turn is proportional to the concentration of the analyte via the stoichiometric relationship of the reaction.

An example is oxidases that produce hydrogen peroxide and water from a substrate (analyte) and oxygen, which can be measured by an amperometer. The concentration of the analyte can be calculated via the stoichiometry of the reaction.

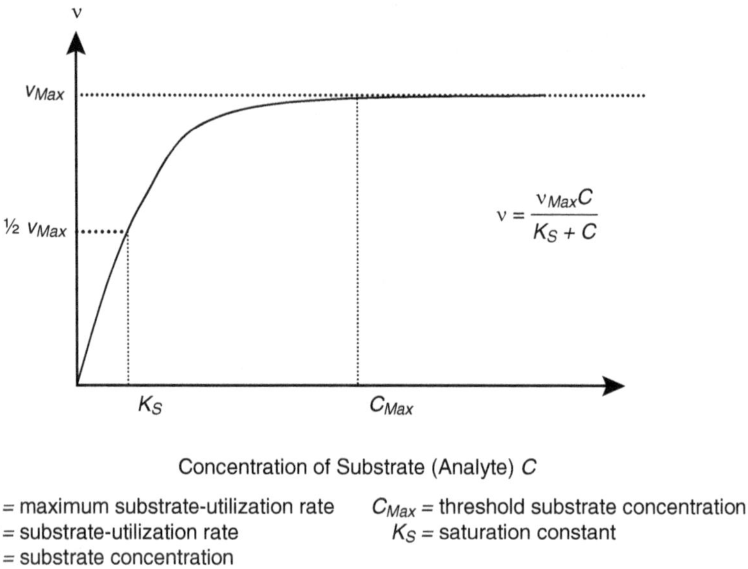

Figure 9-8. Kinetics of enzyme-catalyzed reaction and saturation concentration

The sensitivity of the measurement depends on the specific enzyme reaction, as does the range of the measurement. The kinetics of enzyme-catalyzed reactions follows the Monod Equation, which features a saturation concentration for analyte. Above the saturation concentration, the reaction rate reaches the maximum, as illustrated in Figure 9-8. As a result, the range of analyte is limited up to saturation concentration. Continuous monitoring is possible with repeated calibration. Disposable electrodes may be used in situations where reliability of measurement is emphasized to eliminate the need for repeated calibration.

The basis of conductance measurement is the change in conductance due to the generation or consumption of ionic species from a particular reaction. The ionic strength of the fluid affects the sensitivity. For most water applications (except desalination and ultrapure water treatment), conductance of water is not expected to change significantly. Therefore, the applications of conductance measurement may be limited.

Optical Transducers

Optical transducers convert optical signals into electrical signals. There are many forms of sensing optical signals. Generally, the optic transducer measures either an intensity change or the phase changes that cause light beams to interact or interfere with one another. The former includes light scattering, light adsorption, light excitation (e.g., fluorescence), and light reflectance. Those measured attributes are related to the concentration of the analyte in some manner. The latter (also collectively termed *interferometry*) typically uses an optical device (e.g., semitransparent mirror or a polarizer) to split a light beam into two or more beams and recombine them after letting them travel along different paths. Detection may be either directly related to an analyte or indirectly related to labeled optical tags on an analyte.

The development of fiber optics can be coupled with all the optical transducers and greatly enhanced versatility. The first biosensor intended for environmental monitoring was developed at Oak Ridge National Laboratory by attaching an antibody that reacts specifically with a carcinogen to the end of an optical fiber. Within 5 to 10 minutes after immersing the sensor in groundwater, the antibody/antigen (the carcinogen) reaction produced fluorescence on being illuminated by light at the right wavelength. The fluorescence was transmitted back through the fiber and measured (Jacobson 2007). In another approach, microspheres of various sizes were labeled with antibodies that bind to various bacteria. The sizes of microspheres were identified by their morphological resonances (i.e., the shape-based light emissions when the microspheres were excited by a laser). This is an example of using optical tags for multiple analytes.

Optical transducers have high speed and measurement reproducibility, good sensitivity, and a wide measuring range. They are mostly used for interfacing affinity-based biosensors (e.g., immunosensors) and microorganism-based biosensors. The disadvantages of optical transduction are the need for a sophisticated measuring system and need to hardwire optical transducers to the measuring system for signal transmission.

Acoustic Transducers

Acoustic transducers convert acoustic signals into electrical signals via piezoelectric materials. The change of pressure from sound wave or mass on the surface of a piezoelectric crystal causes the change in its oscillating frequency, which in turn generates an oscillating electrical field or current in a linear fashion. If the change in mass on the crystal surface is very small when compared to the mass of the crystal, the change in oscillating frequency is proportional to the square of the frequency of the crystal oscillation, all else being equal. Therefore, the higher the crystal oscillating frequency, the smaller the detectable change in mass on the crystal surface.

Quartz crystal microbalance is one such acoustic transducer that typically operates at a frequency between 1 and 10 MHz. This corresponds to a detection limit for the change in mass on the crystal surface of 10^{-10} to 10^{-11} g—they are highly sensitive devices. This makes them ideal for coupling with various biological recognition mechanisms by immobilizing enzymes, proteins, antibodies, or microorganisms on the surface of crystal.

Another acoustic transducer is surface acoustic wave (SAW) devices that can operate at up to 250 MHz. However, excessive signal damping in liquid limits their application in water medium. The third type of acoustic transducer is surface transverse wave (i.e., in which the motion of the particle displacement is perpendicular to the direction of wave propagation) transducers. They can operate in liquid even at an oscillating frequency above 250 MHz. Thus, they can provide even higher sensitivity than SAW devices (Rogers and Mascini 2007).

Microcantilevers

In addition to transducers, a relatively recent development in the microcantilever is rapidly progressing for its application in biosensors. A microcantilever—a tiny plate with one end suspended like a miniature diving board—is a simple mechanical device that can act as a physical, chemical, or biological sensor. When a small mass of substances attached to the microcantilever, it bends and changes its vibrating frequencies, which can generates signals directly related to the mass of the substance attached on the microcantilever. Figure 9-9 illustrates microcantilevers in different shapes.

Microcantilevers are typically 100–500 μm long, 20–50 μm wide, and 0.2–1 μm thick and are usually made from silica. On one side of the microcantilever, an active coating is applied for attaching an analyte via various mechanisms. For example, the coating can be gold, enzymes, proteins, DNA, antibodies, or immobilized microorganisms. The interactions between the active coating and an analyte cause the microcantilever to bend (imagine the bending of a diving board when a diver steps on it). The motion of the microcantilever can be transmitted by a wide variety of transduction mechanisms. Depending on whether cantilever deflection or resonance frequency is measured, the measurement can be either static or dynamic (Sepaniak et al. 2002), as illustrated in Figure 9-10.

Static bending of the microcantilever is caused by analyte-induced surface stresses in the different types of coatings. Those stresses can be caused by chemical binding, swelling of a analyte-permeable coating on a surface due to analyte adsorption, or interstitial forces when analytes bind to a nano-structured surface. The changes in the bent microcantilever's radius of curvature, R, and deflection, z_{max}, can be quantitatively related to the stresses by the Stoney Equations as below:

$$\frac{1}{R} = \frac{6(1-\nu)\Delta\sigma}{E\delta} \quad \text{(Eq 9-1)}$$

$$z_{max} = \frac{3L^2(1-\nu)\Delta\sigma}{E\delta^2} \quad \text{(Eq 9-2)}$$

Where

R = radius of curvature

ν = Poisson's ratio

E = Yong's modulus for the substrate

$\Delta\sigma$ = differential surface stress

δ = thickness of microcantilever

L = length of microcantilever

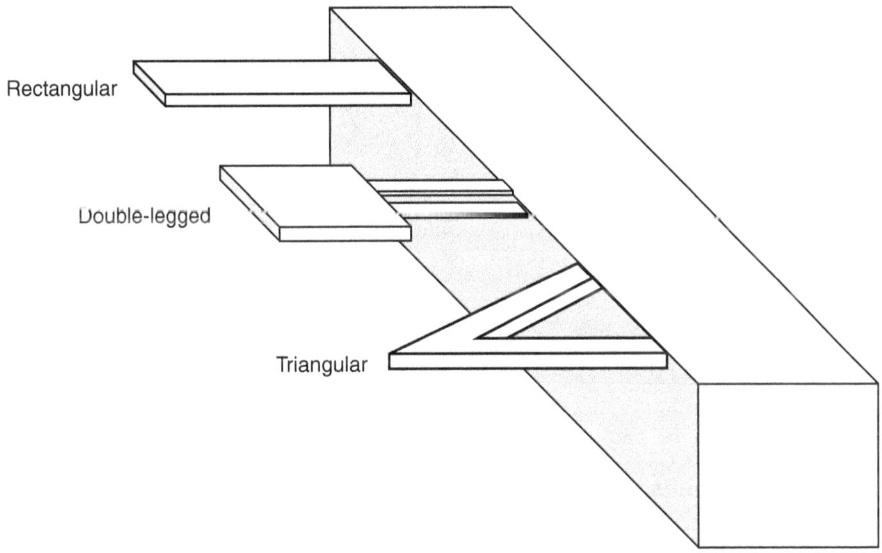

Figure 9-9. Microcantilevers in various shapes

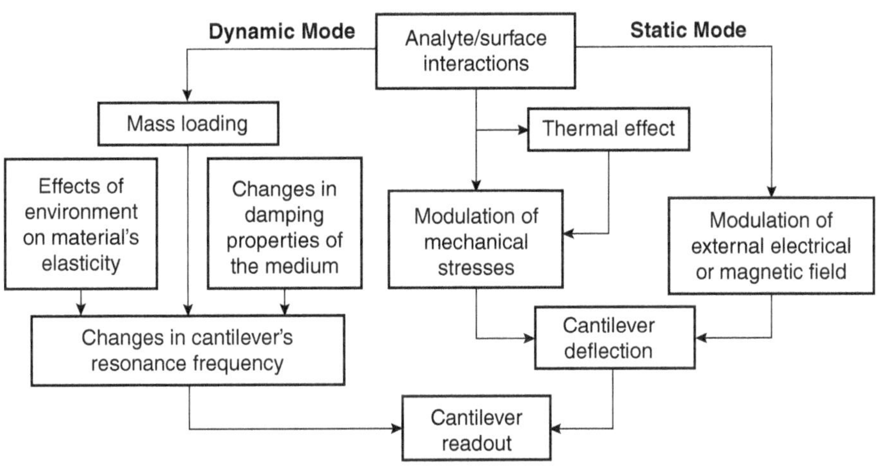

Figure 9-10. Transduction principles of microcantilever (adapted from Sepaniak et al. 2002)

The deflection of the microcantilever can be detected via piezoresistive or optical means. The piezoresistive method involves embedding piezoresistive material near the top of the microcantilever to record the stress change occurring at its surface. The piezoresistive microcantilever can be incorporated into one leg of the Wheatstone bridge circuit as illustrated in Figure 9-11, in which deflection of the microcantilever causes a change in the resistance in the leg and so a change to the voltage across the bridge.

The optical detection of the deflection of the microcantilever is illustrated in Figure 9-12. An ultralow-power laser beam is aimed at the microcantilever and bounced

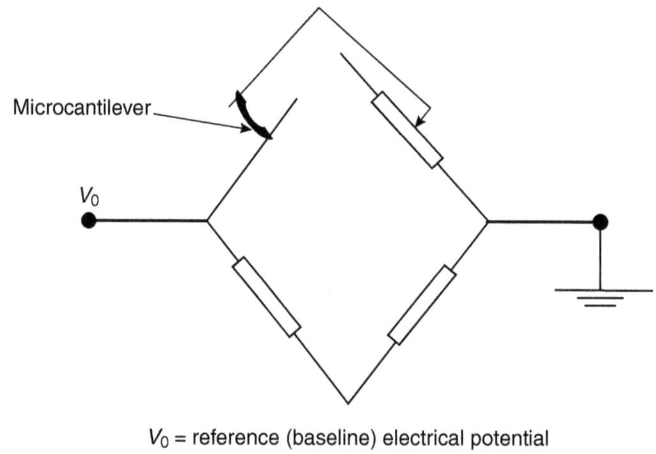

Figure 9-11. Piezoresistive detection of microcantilever deflection by a Wheatstone bridge circuit with an incorporated piezoresistive microcantilever

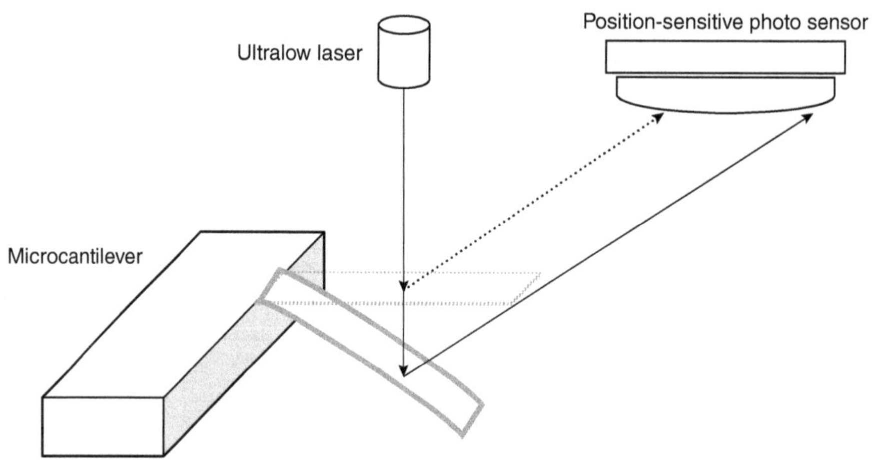

Figure 9-12. Optical detection of microcantilever deflection

off by the reflective coating to a position-sensitive photo sensor. When the microcantilever is deflected, it changes the position of the reflected laser beam and this change is registered by the position-sensitive photo sensor.

In dynamic mode, a microcantilever acts similarly to an acoustic oscillator. The change in mass caused by the chemical binding or affinity adsorption changes the resonance frequency as described by the following equation:

$$\frac{1}{f_1^2} - \frac{1}{f_0^2} = \frac{\Delta m}{4\pi^2 K} \qquad \text{(Eq 9-3)}$$

Where

f_1, f_0 = resonance frequencies with and without analyte, respectively

Δm = change in mass

K = spring constant of microcantilever

Compared with acoustic transducers such as quartz crystal microbalance and surface acoustic wave devices, the microcantilever exhibits sensitivity that is orders of magnitude higher. This high sensitivity is the result of low stiffness—a structural feature of microcantilevers. For a nano-scale cantilever with a fundamental frequency in the range of 10^5–10^6 Hz, the calculated mass sensitivity is down to 10^{-21} g—approximately the mass of a single molecule (Sepaniak et. al. 2002). In reality, this mass sensitivity would be limited by the damping effect of the medium. Even so, such a sensitivity is still remarkably high and capable of detecting a single bacterium (a mass of approximately 10^{-17} to 10^{-18} g).

The earlier application of microcantilevers as chemical sensors was reported for measuring mercury vapor at the level of 10^{-12} g (Thundat et al. 1995) and vapors of aromatic compounds at the concentration of subparts per billion (Ji et al. 2000; Ji et al. 2001). Coating a microcantilever with biological receptors enables it to be used as a biosensor. Hansen et al. reported that the oligonucleotide hybridization was able to detect a single mismatch of a base pair on a DNA sequence (Hansen et al. 2001). Microcantilever biosensors provide a label-free version of the immunoassay for detecting a prostate-specific antigen down to 10^{-10} g/mL (Wu et al. 2001).

Application of the microcantilever in the detection of environmental contaminants is still in its infancy. Issues yet to be solved for it applications in environmental monitoring include nonspecific binding of surfaces and the damping effect of liquid. Background noises such as flow and pressure fluctuations, external electrical and magnetic fields, and mechanical vibrations all can affect the measurement. In addition, how to maintain long-term stability of biological components for recognition also needs to be evaluated. Nevertheless, the high sensitivity, real-time detection, and miniaturization capacity of the microcantilever-based biosensor point to a promising future.

CONCLUSION

For membrane integrity monitoring, we have established that sensitivity, reliability, continuity, and traceability are very important issues. An ideal device for monitoring membrane integrity would be able not only to detect a very low level of targeted contaminant due to a membrane breach reliably and in real time but also to identify to the plant operator where the membrane breach is. This would require that a sensor not only senses but also reports a membrane integrity breach. This is particularly important for membrane plants that have many membrane modules. To meet the reporting requirement sensors would need to be combined with transmitters into one unit and arranged in a distributed manner. Multiplexing discussed in chapter 6 is an example of such a setup, in which the task is accomplished by running optical fibers from the central unit to individual sensors. Sensors that can transmit signals via wireless transmitters may also be used for this purpose. With advances in technologies for biosensors, computing, and communications and the convergence of those technologies, it may not be difficult to envision that one day sensors will not only alert the plant operator that there is a membrane integrity breach but also be able to "broadcast" the location of such a breach.

Appendix A

METRIC CONVERSION FACTORS (US CUSTOMARY UNITS TO SI UNITS)

Multiply the US customary unit			To obtain the SI unit	
Name	Symbol	By	Symbol	Name
Length				
feet	ft	0.3048	m	meter
inch	in.	25.4	mm	millimeter
		2.54×10^4	μm	micrometer
		2.54×10^7	nm	nanometer
		2.54×10^8	Å	Angstrom unit
Area				
square feet	ft^2	0.0929	m^2	square meter
square inch	in^2	6.452	cm^2	square centimeter
Volume				
cubic feet	ft^3	0.0283	m^3	cubic meter
		28.3	L	liter
gallon	gal	3.785	L	liter
		3,785	mL	milliliter
cubic inch	in^3	16.387	mL	milliliter
		1.64×10^4	μL	microliter
Velocity				
feet per second	ft/s	0.3048	m/s	meters per second
Acceleration				
feet per second squared	ft/s^2	0.3048	m/s^2	meters per second squared
Flowrate				
cubic feet per second	ft^3/s	0.0283	m3/s	cubic meters per second
		28.3	L/s	liters per second
million gallons per day	mgd	3.785	mld	million liters per day
		3,785	m^3/day	cubic meters per day
gallons per minute	gpm	3.785	L/min	liters per minute
		3.79×10^3	mL/min	milliliters per minute

Multiply the US customary unit			To obtain the SI unit	
Name	**Symbol**	**By**	**Symbol**	**Name**
Flux				
gallons per square feet per minute	gfd		lmh	liters per hour per square meters
Mass				
pound	lb	0.454	kg	kilogram
		454	g	gram
		4.54×10^5	mg	milligram
		4.54×10^8	µg	microgram
		4.54×10^{11}	Ng	nanogram
Force				
pound force	lb_f	4.45×10^{-3}	kN	kilonewton
		4.448	N	Newton
Pressure				
atmosphere	atm	0.1013	MPa	megapascals
		101.3	kPa	kilopascals
		10	m H_2O	meters of water
feet of water	ft H_2O	0.3048	m H_2O	meters of water
pound (force) square inch	psi	6.894	kPa	kilopascals
		6,894	Pa	pascals (newtons per square meter)
Temperature				
degrees Fahrenheit	°F			
		0.555(°F – 32)	°C	degrees Celsius (centigrade)
degrees Rankine	°R	0.555(°F + 459.67)	°K	degrees Kelvin
		0.555(°R – 491.67)	°C	degrees Celsius (centigrade)
		0.555	°K	degrees Kelvin
Power				
foot-pounds (force) per second	ft·lbf/s	1.36×10^{-3}	kW	kilowatts
horsepower	hp	1.3558	W	watts
		0.7457	kW	kilowatts
Energy				
foot-pounds (force)	ft·lb_f	1.3558	J	joules
horsepower-hour	hp·h	2.6845	MJ	megajoules
kilowatt-hour	kWh	3600	kJ	kilojoules

Appendix B

PHYSICAL PROPERTIES OF WATER

U.S. Customary Units

Temperature °F	Specific Weight γ lb/ft^3	Density ρ slug/ft^3	Dynamic Density $\mu \times 10^5$ lb·s/ft^2	Kinematic Viscosity $\nu \times 10^5$ ft^2/s	Surface Tension $\sigma \times 10^3$ lb/ft	Vapor Pressure p_v psi
32	62.42	1.940	3.746	1.931	5.18	0.09
40	62.43	1.940	3.229	1.664	6.14	0.12
50	62.41	1.940	2.735	1.410	5.09	0.18
60	62.37	1.938	2.359	1.217	5.04	0.26
70	62.30	1.936	2.050	1.059	4.98	0.36
80	62.22	1.934	1.799	0.930	4.92	0.51
90	62.11	1.931	1.595	0.826	4.86	0.70
100	62.00	1.927	1.424	0.739	4.80	0.95
110	61.86	1.923	1.284	0.667	4.73	1.27
120	61.71	1.918	1.168	0.609	4.67	1.69
130	61.55	1.913	1.069	0.558	4.60	2.22
140	61.38	1.908	0.981	0.514	4.54	2.89
150	61.20	1.902	0.905	0.476	4.47	3.72
160	61.00	1.896	0.838	0.422	4.41	4.74
170	60.80	1.890	0.780	0.413	4.34	5.99
180	60.58	1.883	0.726	0.385	4.27	7.51
190	60.36	1.876	0.678	0.362	4.20	9.34
200	60.12	1.868	0.637	0.341	4.13	11.52
212	59.83	1.860	0.593	0.319	4.04	14.70

NOTE: 1 slug = 1 lb$_f$s^2/ft

SI Units

Temperature °C	Specific Weight γ kN/m^3	Density ρ × 10–3 kg/m^3	Dynamic Density μ × 103 N·s/m^2	Kinematic Viscosity ν × 106 m^2/s	Surface Tension σ × 102 N/m	Vapor Pressure p_v kN/m^2
0	9.805	0.9998	1.781	1.785	7.65	0.61
5	9.807	1.0000	1.518	1.519	7.49	0.87
10	9.804	0.9997	1.307	1.306	7.42	1.23
15	9.798	0.9991	1.139	1.139	7.35	1.70
20	9.789	0.9982	1.002	1.003	7.28	2.34
25	9.777	0.9970	0.890	0.893	7.20	3.17
30	9.764	0.9957	0.798	0.800	7.12	4.24
40	9.730	0.9922	0.653	0.658	6.96	7.38
50	9.689	0.9880	0.547	0.553	6.79	12.33
60	9.642	0.9832	0.466	0.474	6.62	19.92
70	9.589	0.9778	0.404	0.413	6.44	31.16
80	9.530	0.9718	0.354	0.364	6.26	47.34
90	9.466	0.9653	0.315	0.326	6.08	70.10
100	9.399	0.9584	0.282	0.294	5.89	101.39

Source: Adapted from *Wastewater Engineering: Treatment, Disposal, and Reuse*, 3rd ed.

Appendix C

Excerpts From USEPA's Membrane Filtration Guidance Manual (MFGM) Chapter 4 and Appendices C and D

4.0 Direct Integrity Testing

4.1 Introduction

In order for a membrane process to be an effective barrier against pathogens and other particulate matter, the filtration system must be integral, or free of any integrity breaches. Thus, it is critical that operators are able to demonstrate the integrity of this barrier on an ongoing basis during system operation. Direct integrity testing represents the most accurate means of assessing the integrity of a membrane filtration system that is currently available.

Under the Long Term 2 Enhanced Surface Water Treatment Rule (LT2ESWTR), a direct integrity test is defined as a physical test applied to a membrane unit in order to identify and isolate integrity breaches. In order to receive *Cryptosporidium* removal credit for compliance with the rule, the removal efficiency of a membrane filtration process must be routinely verified during operation using direct integrity testing. The direct integrity test must be applied to the physical elements of the entire membrane unit, including membranes, seals, potting material, associated valves and piping, and all other components which could result in contamination of the filtrate under compromised conditions (40 CFR 141.719(b)(3)(i)).

There are two general classes of direct integrity tests that are commonly used in membrane filtration facilities: ***pressure-based tests*** and ***marker-based tests***. The pressure-based tests are based on bubble point theory (as described in Appendix B) and involve applying a pressure or vacuum (i.e., negative pressure) to one side of a membrane barrier and monitoring for parameters such as pressure loss or the displacement of air or water in order to establish whether an integrity breach is present. The various pressure-based tests include the pressure and vacuum decay tests, the diffusive airflow test, and the water displacement test. Marker-based tests utilize either a spiked particulate or molecular marker to verify membrane integrity by directly assessing removal of the marker, similar to a challenge test.

The LT2ESWTR does not require the use of a particular direct integrity test for rule compliance, but rather that any test used meet the specified performance criteria for ***resolution***, ***sensitivity***, and ***frequency***. Thus, a particular system may utilize an appropriate pressure- or marker-based test or any other method that both meets the performance criteria and is approved by the State. The performance criteria for direct integrity tests are summarized as follows:

- **Resolution**: The direct integrity test must be responsive to an integrity breach on the order of 3 μm or less (40 CFR 141.719(b)(3)(ii)).

- **Sensitivity**: The direct integrity test must be able to verify a log removal value (LRV) equal to or greater than the removal credit awarded to the membrane filtration process (40 CFR 141.719(b)(3)(iii)).

- **Frequency**: A direct integrity test must be conducted on ***each membrane unit*** at a frequency of no less than once each day that the unit is in operation. (The definition of a membrane unit under the LT2ESWTR is provided in section 1.5.) Less frequent testing may be approved by the State if supported by demonstrated process reliability, the use of multiple barriers effective for *Cryptosporidium*, or reliable process safeguards (40 CFR 141.719(b)(3)(vi)).

In addition to the performance criteria, the rule also requires the establishment of a ***control limit*** within the associated sensitivity limits for the direct integrity test that is indicative of an integral membrane unit capable of achieving the *Cryptosporidium* removal credit awarded by the State (40 CFR 141.719(b)(3)(iv)). If the results of the direct integrity test exceed this limit, the rule requires that the affected membrane unit be taken off-line for diagnostic testing and repair (40 CFR 141.719(b)(3)(v)). The unit may be returned to service only after system integrity has been confirmed by a direct integrity test.

The objective of Chapter 4 is to describe the various pressure- and marker-based direct integrity tests currently in use and how these tests can be applied to meet the performance criteria specified under the LT2ESWTR. Diagnostic tests, data collection, and reporting are also addressed.

This chapter is divided into the following sections:

Section 4.2: Test Resolution
This section discusses the determination of pressure- and marker-based direct integrity test resolution for meeting the performance criteria required by the rule.

Section 4.3: Test Sensitivity
This section discusses the determination of pressure- and marker-based direct integrity test sensitivity for meeting the performance criteria required by the rule, including general concepts and methods.

Section 4.4: Test Frequency
This section reviews the direct integrity testing frequency requirements of the rule.

Section 4.5: Establishing Control Limits
This section describes the mathematical and experimental determination of control limits for direct integrity testing.

Section 4.6: Example: Establishing Direct Integrity Test Parameters
This section illustrates the calculation of some of the critical direct integrity test performance criteria, including test resolution, sensitivity, and control limits for an example membrane filtration system.

Section 4.7: Test Methods
This section provides an overview of the various types of pressure- and marker-based tests, including generic test protocols as well as some advantages and disadvantages of each.

Section 4.8: Diagnostic Testing
This section describes some of the diagnostic tests that are used to identify and isolate integrity breaches following a failed direct integrity test.

Section 4.9: Data Collection and Reporting
This section provides guidance on direct integrity test data collection and reviews the associated reporting requirements of the rule.

4.2 Test Resolution

Resolution is defined as the size of the smallest integrity breach that contributes to a response from a direct integrity test. Any direct integrity test applied to meet the requirements of the LT2ESWTR is required to have a resolution of 3 μm or less. This resolution criterion is based on the lower size range of *Cryptosporidium* oocysts and is intended to ensure that any integrity breach large enough to pass oocysts contributes to a response from the direct integrity test used. The manner in which the resolution criterion is met depends on whether the direct integrity test is pressure-based or marker-based, as described in the following subsections.

4.2.1 Pressure-Based Tests

In order to achieve a resolution of 3 μm with pressure-based direct integrity tests, the net pressure applied during the test must be great enough to overcome the capillary forces in a 3 μm hole, thus ensuring that any breach large enough to pass *Cryptosporidium* oocysts would also pass air during the test. The minimum applied test pressure necessary to achieve a resolution of 3 μm is calculated using Equation 4.1:

$$P_{test} = (0.193 \bullet \kappa \bullet \sigma \bullet \cos\theta) + BP_{max} \qquad \text{Equation 4.1}$$

Where:

P_{test}	=	minimum test pressure (psi)
κ	=	pore shape correction factor (dimensionless)
σ	=	surface tension at the air-liquid interface (dynes/cm)
θ	=	liquid-membrane contact angle (degrees)
BP_{max}	=	maximum backpressure on the system during the test (psi)
0.193	=	constant that includes the defect diameter (i.e., 3 μm resolution requirement) and unit conversion factors

Equation 4.1 is based on bubble point theory and is derived from the balancing of capillary static forces. Note that the constant of 0.193 accounts for the LT2ESWTR resolution requirement of responding to a defect of at least 3 μm in diameter, as well as the appropriate unit conversion factors, in order to simplify the equation for the purposes of rule compliance. The general form of Equation 4.1 includes the capillary diameter as a variable and represents an expression relating this diameter to the bubble point pressure. A discussion of bubble point theory and a derivation of Equation 4.1 is provided in Appendix B.

Values for several parameters in Equation 4.1 must be determined in order to calculate the minimum test pressure necessary to achieve a resolution of 3 μm when using a pressure-based integrity test. The parameters κ and θ are intrinsic properties of the membrane. In the absence of data supplied by the membrane manufacturer, conservative values of $\kappa = 1$ and $\theta = 0$ should be used. Appendix B provides an additional discussion of these parameters. The surface tension, σ, is inversely related to temperature; consequently, the surface tension at the coldest anticipated water temperature should be used to calculate a conservative value for the minimum required test pressure. As a point of reference, the surface tension of water at 5 °C is 74.9 dynes/cm. Substituting these three values (i.e., $\kappa = 1$, $\theta = 0$, and $\sigma = 74.9$ dynes/cm) into Equation 4.1 yields the following simplified equation:

$$P_{test} = 14.5 + BP_{\max} \qquad \text{Equation 4.2}$$

Where:
- P_{test} = minimum test pressure (psi)
- BP_{max} = maximum backpressure on the system during the test (psi)

Equation 4.2 indicates that the minimum test pressure necessary to achieve a 3-μm resolution is 14.5 psi plus the maximum backpressure on the system during application of a pressure-based direct integrity test (at a conservative temperature of 5 °C). Ideally, there should be no hydrostatic backpressure on the system during the test. However, it is not always practical to perform the test without any hydrostatic backpressure, and in these cases the additional backpressure must be considered in establishing the minimum test pressure necessary to meet the resolution criterion. For example, there might be hydrostatic pressure on the undrained side of the membrane if a pressure-driven membrane module remains filled with water or if a vacuum-driven (i.e., immersed) membrane remains submerged under water in a basin. Thus, if the bottom of the membrane is under 7 feet of water, BP_{max} would be approximately 3 psi, yielding a P_{test} value of 17.5 psi to achieve a resolution of 3 μm.

Both Equations 4.1 and 4.2 assume that the applied pressure remains constant during the direct integrity test. However, in many cases there may be some baseline decay (e.g., that attributable to diffusion) that is measurable over the duration of the test. In this case, it is important to account for this baseline decay in the resolution calculation. Thus, in order to ensure that the resolution requirement is satisfied throughout the duration of the test, the anticipated pressure at the end of the direct integrity test should be used to calculate the resolution. This value can be estimated using the initial applied pressure, the typical rate of baseline pressure decay for a fully integral system, and the duration of the test. If the baseline decay is small enough such that the final test pressure is within approximately 5 percent of the initial applied pressure, the baseline decay can be assumed to be negligible, and the initial applied pressure may be used to calculate the test resolution.

The LT2ESWTR does not establish the minimum test pressure to be used during a pressure-based direct integrity test, but rather only requires that the test achieve a 3-μm resolution. If a membrane manufacturer has information to support the use of values other than $\kappa = 1$ and $\theta = 0$, and these less conservative values are approved by the State, then Equation 4.1 can be used to calculate the minimum required test pressure. It is essential that the use of values other than $\kappa = 1$ and $\theta = 0$ be scientifically defensible, since the use of inappropriate values could result in the use of a test pressure that does not meet the resolution criterion established by the rule. One approach for determining membrane-specific values for κ and θ is through direct experimental evaluation. Because these parameters can have a significant effect on the required direct integrity test pressure, it is strongly recommended that States require sufficient justification from a membrane manufacturer prior to approving the use of values other than $\kappa = 1$ and $\theta = 0$, such as independent third party testing results using a method accepted by the scientific community and demonstrating statistically significant data.

Although the rule does not include a frequency requirement for the recalculation of resolution, the resolution should be recalculated if the system backpressure during direct integrity testing is adjusted. Alternatively, if desired, direct integrity test instrumentation and the data recording system could be configured to calculate the resolution after each application of the direct integrity test using the applied test pressure, system backpressure, and surface tension corresponding to the temperature at which the test is conducted. Note that the liquid-membrane contact angle can also change over the life of a membrane module (e.g., as a result of the adsorption of organic matter by the membrane material), and these changes may not necessarily be uniform among the various modules in a unit (Childress et al. 1996; Jucker et al. 1994). Thus, if a value other than $\theta = 0$ (i.e., the most conservative value) is used, then it may be appropriate to periodically recalculate the resolution based on a revised estimate of the actual value of θ, an exercise that may necessitate destructive testing of a representative sample of membrane modules in the system.

4.2.2 Marker-Based Tests

A marker-based direct integrity test can be viewed as a "mini challenge study," in which a surrogate is periodically applied to the feed water in order to verify the integrity of a membrane filtration system. In order to meet the resolution criterion of the rule, the surrogate used in a marker-based test must have an effective size of 3 μm or smaller, as described in section 3.9.2. A marker-based direct integrity test can use either particulate or molecular surrogates, but in either case, it must be established that the surrogate meets the resolution criterion. Section 3.9 presents guidelines for the selection of a conservative surrogate for *Cryptosporidium* during challenge testing, and these same guidelines are applicable to selection of an appropriate surrogate for a marker-based direct integrity test. The effective size of the marker can be established through any accepted methodology such as size distribution analysis of particulate markers or estimation techniques based on the molecular weight and geometry of molecular markers.

4.3 Test Sensitivity

Sensitivity is defined as the maximum log removal value that can be reliably verified by the direct integrity test (i.e., LRV_{DIT}). The sensitivity of the direct integrity test establishes the maximum log removal credit that a membrane filtration process is eligible to receive if it is less than or equal to that demonstrated during challenge testing (i.e., $LRV_{C\text{-}Test}$). For example, if the challenge test demonstrated a $LRV_{C\text{-}Test}$ of 5.5 log, and the direct integrity test is capable of verifying a LRV_{DIT} of 4.5 log, the membrane filtration process would be eligible for removal credit up to 4.5 log. Although the sensitivity of the direct integrity test should not be expected to vary significantly over time, the determination of sensitivity as described in this section is designed to produce a conservative result that would remain applicable over the life of the membrane filtration system. However, if significant changes occur in terms of operational parameters, direct integrity test conditions, or any basic assumptions that might affect the value of the direct integrity test sensitivity, it is suggested that the sensitivity be reestablished to verify that it is at least equal to the removal credit awarded to the process.

The sensitivity of a direct integrity test is logarithmic in nature. For example, a test with a LRV_{DIT} of 5 log is 100 times more sensitive than a test with a LRV_{DIT} of 3 log. Thus, when a higher sensitivity is required, the test must be capable of measuring very small changes in the direct integrity test response and distinguishing these results from background or baseline data. Data suggest that many direct integrity tests, as currently applied, have sensitivities in excess of 4 log; however, sensitivity must be determined on a case-by-case basis using the information provided by the membrane manufacturer and the guidance in this document. While determination of integrity test sensitivity can be complex, it provides a rational basis for awarding high removal credits to membrane processes that are commensurate with their abilities. As was the case with resolution, the manner in which sensitivity is determined depends on whether the type of direct integrity test used is pressure- or marker-based.

Note that for systems that utilize multiple stages of membrane filtration, the sensitivity for each stage must be determined independently. The most common example of such an application would be a case in which a second stage is used to treat the backwash water from the first stage, after which the filtrate from the two stages is blended (i.e., concentrate staging). In this case, as well as in others involving multiple stages, if the filtrate from the various stages is blended, the stage using the membrane units with the lowest sensitivity would limit the maximum log removal credit that could be awarded to the overall process. However, if a second (or subsequent) stage is used strictly for residuals management such that the filtrate is recycled to an upstream point in the overall treatment process, the LT2ESWTR would not be applicable to such ancillary stages.

4.3.1 Pressure-Based Tests

The discussion in this section regarding the calculation of sensitivity for pressure-based direct integrity tests is divided into three parts, as follow. First, the basic concepts that are applicable to all pressure-based tests are introduced. The subsequent section describes calculation of the sensitivity for pressure-based direct integrity tests based on this general conceptual framework. A third section discusses the determination of diffusive (or baseline) losses in a fully integral system during the application of a pressure-based direct integrity test.

4.3.1.1 Basic Concepts

The determination of sensitivity for pressure-based direct integrity tests is more complex than for marker-based tests. The equation used to determine the sensitivity of a pressure-based direct integrity test is specified in the LT2ESWTR (40 CFR 141.719(b)(3)(iii)(A)) and given below as Equation 4.3:

$$LRV_{DIT} = \log\left(\frac{Q_p}{VCF \bullet Q_{breach}}\right) \qquad \text{Equation 4.3}$$

Where:

LRV_{DIT}	=	direct integrity test sensitivity in terms of LRV (dimensionless)
Q_p	=	membrane unit design capacity filtrate flow (L/min)
Q_{breach}	=	flow from the breach associated with the smallest integrity test response that can be reliably measured, referred to as the *critical breach size* (L/min)
VCF	=	volumetric concentration factor (dimensionless)

Equation 4.3 represents a dilution model that assumes water passing through the intact membrane barrier is free of the particulate contaminant of interest and that water flowing through an integrity breach has a particulate contaminant concentration equal to that on the high pressure side of the membrane. Under these assumptions, LRV_{DIT} is a function of the ratio of total filtrate flow to flow through the critical breach (i.e., Q_p/Q_{breach}), which quantifies the dilution of the contaminated stream passing through the breach as it mixes with treated filtrate. For a membrane unit of a given capacity (i.e., constant Q_p), LRV_{DIT} will increase as Q_{breach} decreases. This implies that a more sensitive direct integrity test capable of detecting a smaller breach can verify a higher log removal value and thus potentially increase the removal credit that a membrane filtration system is eligible to receive.

The volumetric concentration factor (VCF) is a dimensionless term that accounts for the increase in the suspended solids concentration that occurs on the feed side of the membrane for some hydraulic configurations. The VCF is important in the context of membrane integrity because the risk of filtrate contamination in the event of an integrity breach is increased for systems in which the influent suspended solids are concentrated on the feed side of the membrane. For example, for the same size integrity breach, systems with higher VCFs will allow increased passage of pathogens to the filtrate, reducing the verifiable removal efficiency. This affect is taken into account in the determination of sensitivity for pressure-based direct integrity tests in that systems with higher VCFs have proportionately lower test sensitivity. Thus, in the denominator of Equation 4.3, the term Q_{breach} is multiplied by the VCF to account for the impact of this concentration effect.

The VCF is calculated as the ratio of the concentration of suspended solids maintained on the feed side of the membrane to that in the influent feed water, as shown in Equation 4.4 (also Equation 2.22):

$$VCF = \frac{C_m}{C_f} \qquad \text{Equation 4.4 (2.22)}$$

Where:

VCF	=	volumetric concentration factor (dimensionless)
C_m	=	concentration of suspended solids maintained on the feed side of the membrane (number or mass / volume)
C_f	=	concentration of suspended solids in the influent feed water to the membrane system (number or mass / volume)

The VCF generally ranges between 1 and 20, and the value depends on the hydraulic configuration of the system. Membrane systems that operate in deposition mode do not increase the concentration of suspended solids on the feed side of the membrane and thus have a VCF equal to one. In contrast, membrane systems that operate in suspension mode, such as those that utilize a crossflow hydraulic configuration, typically have a VCF in the range of 4 to 20, representing a 4- to 20-fold increase in the suspended particle concentration on the feed side of the membrane. The methods and equations used to calculate the VCF for various hydraulic configurations are provided in section 2.5, and Table 2.4 presents equations for calculating both the average and maximum VCF for various hydraulic configurations. Alternatively, the VCF could be determined experimentally, as discussed in section 2.5.4.

Note that the LT2ESWTR does not specify use of the maximum or average VCF value in calculating sensitivity, but the rule does require that the increase in suspended solids concentration on the high-pressure side of the membrane, as occurs with some hydraulic configurations, be considered in the calculation. The maximum VCF typically ranges from 1 to 20 and provides the most conservative value for LRV_{DIT}, while the average VCF typically ranges from 1 to 7. In selecting between the maximum, average, or any other value for the VCF, consideration should be given to the concentration profile along the membrane surface in the direction of water flow and the implication of integrity breaches at various locations within the membrane unit. For example, although the maximum VCF does provide the most conservative value for LRV_{DIT}, for systems in which the concentration varies with position this value represents only a very small portion of the concentration profile and thus is only representative of breaches that occur at the extreme end of the membrane unit. Similarly, some systems exhibit a concentration profile as a function of time within a filtration cycle, and the maximum VCF only occurs at the end of the filtration cycle immediately before a backwash event; prior to this time, the VCF is significantly lower.

4.3.1.2 Calculating Sensitivity

The sensitivity of a pressure-based direct integrity test can be calculated by converting the response from a pressure-based test that measures the flow of air (e.g., the diffusive airflow test) or rate of pressure loss (e.g., the pressure decay test) to an equivalent flow of water through an integrity breach during normal operation, as described in the following subsection. The second subsection outlines a general procedure for determining the threshold response of a pressure-based direct integrity test experimentally, if this information is not available from the membrane filtration system manufacturer.

<u>Calculating Sensitivity Using the Air-Liquid Conversion Ratio</u>

In order to calculate the LRV_{DIT}, the flow through the critical breach for a direct integrity test (i.e., Q_{breach}) must be determined (as shown in Equation 4.3). Since most direct integrity tests do not directly measure Q_{breach}, it is necessary to establish a correlation between the direct integrity test response and the flow of water through the critical breach during system operation. In some of the most commonly used pressure-based direct integrity tests, including the pressure- or vacuum-decay test and the diffusive airflow test, air is applied to the drained side of a membrane and subsequently flows through any integrity breach that exceeds the test resolution. The response from such a test is typically measured as pressure decay or airflow. In order to relate the response from a pressure-based integrity test to Q_{breach}, it is necessary to establish a correlation between airflow and liquid flow through the critical integrity breach.

This correlation can be characterized through the air-liquid conversion ratio (ALCR), which is defined as the ratio of air that would flow through a breach during a direct integrity test to the amount of water that would flow through the breach during filtration, as defined in Equation 4.5:

$$ALCR = \left(\frac{Q_{air}}{Q_{breach}}\right) \qquad \text{Equation 4.5}$$

Where:

	ALCR	=	air-liquid conversion ratio (dimensionless)
	Q_{air}	=	flow of air through the critical breach during a pressure-based direct integrity test (L/min)
	Q_{breach}	=	flow of water through the critical breach during filtration (L/min)

The ALCR can be used to express the liquid flow through a breach in terms of corresponding flow of air, as shown in Equation 4.6:

$$Q_{breach} = \left(\frac{Q_{air}}{ALCR} \right) \qquad \text{Equation 4.6}$$

Where:
- Q_{breach} = flow of water through the critical breach during filtration (L/min)
- Q_{air} = flow of air through the critical breach during a pressure-based direct integrity test (L/min)
- ALCR = air-liquid conversion ratio (dimensionless)

Substituting Equation 4.6 into the general expression for sensitivity (Equation 4.3) yields the following expression:

$$LRV_{DIT} = \log\left(\frac{Q_p \bullet ALCR}{Q_{air} \bullet VCF} \right) \qquad \text{Equation 4.7}$$

Where:
- LRV_{DIT} = direct integrity test sensitivity in terms of LRV (dimensionless)
- Q_p = membrane unit design capacity filtrate flow (L/min)
- ALCR = air-liquid conversion ratio (dimensionless)
- Q_{air} = flow of air through the critical breach during a pressure-based direct integrity test (L/min)
- VCF = volumetric concentration factor (dimensionless)

Equation 4.7 can be used to directly calculate the sensitivity for any pressure-based direct integrity test that is based on bubble point theory and which measures the flow of air (Q_{air}) through an integrity breach. The four parameters that need to be determined to calculate sensitivity are: Q_p, VCF, ALCR, and Q_{air}. The VCF can be established as previously described in section 4.3.1.1. Q_p is the design capacity filtrate flow approved by the State from the membrane unit to which the direct integrity test is applied. For a constant Q_{breach}, higher filtrate flows yield greater direct integrity test sensitivity. Thus, if different sizes of membrane units with varying capacity are used in the same treatment system, the sensitivity of the direct integrity test should be independently determined for each unit size, with the lowest sensitivity establishing the maximum log removal credit for the overall process. The flow of air, Q_{air}, is related to the response from the direct integrity test. For tests that measure the airflow through an integrity breach directly, such as the diffusive airflow test, Q_{air} is simply the airflow measured during application of the test. On the other hand, methods such as the pressure and vacuum decay tests yield results in terms of pressure loss per unit time, which must be converted to an equivalent flow of air using Equation 4.8:

$$Q_{air} = \frac{\Delta P_{test} \bullet V_{sys}}{P_{atm}} \qquad \text{Equation 4.8}$$

Where:

Q_{air}	=	flow of air (L/min)
ΔP_{test}	=	rate of pressure decay during the integrity test (psi/min)
V_{sys}	=	volume of pressurized air in the system during the test (L)
P_{atm}	=	atmospheric pressure (psia)

Note that Equation 4.8 assumes that the temperature of the water and air are the same since the air temperature should rapidly equilibrate with the water temperature. V_{sys} encompasses the entire pressurized volume, including all fibers (typically the insides of the fibers are pressurized), piping, and other void space on the pressurized side.

Substituting Equation 4.8 into Equation 4.7 yields Equation 4.9, which can be used to calculate the sensitivity of a direct integrity test that measures the rate of pressure or vacuum decay.

$$LRV_{DIT} = \log\left(\frac{Q_p \bullet ALCR \bullet P_{atm}}{\Delta P_{test} \bullet V_{sys} \bullet VCF}\right) \qquad \text{Equation 4.9}$$

Where:

LRV_{DIT}	=	direct integrity test sensitivity in terms of LRV (dimensionless)
Q_p	=	membrane unit design capacity filtrate flow (L/min)
ALCR	=	air-liquid conversion ratio (dimensionless)
P_{atm}	=	atmospheric pressure (psia)
ΔP_{test}	=	smallest rate of pressure decay that can be reliably measured and associated with a known integrity breach during the integrity test (psi/min)
V_{sys}	=	volume of pressurized air in the system during the test (L)
VCF	=	volumetric concentration factor (dimensionless)

Regardless of whether the flow of air (Q_{air}) or pressure decay rate (ΔP_{test}) is measured during the direct integrity test, the smallest response from the test that can be reliably measured and associated with an integrity breach should be used in the sensitivity calculation. This should not be confused with the baseline integrity test response from an ***integral*** membrane unit, since there may be a small airflow or pressure decay due to diffusion of air through water in the wetted pores and/or the membrane material, even if there are no breaches in the system. In many cases, this smallest measurable response associated with an integrity breach may be provided by the membrane filtration system manufacturer. If this information is not available from the system manufacturer, it may be determined experimentally by progressively creating small integrity breaches of a known size in an otherwise integral membrane unit in order to determine the smallest measurable response from the direct integrity test that is distinguishable from the baseline response for an integral membrane unit. A general procedure for this experimental method is described in the following subsection ("Measuring The Threshold Direct Integrity Test Response Experimentally").

The basic procedure for calculating the ALCR involves first making a reasonable assumption regarding whether the flow through the critical breach is laminar or turbulent for the particular membrane filtration system of interest. Then the ALCR can be calculated using the appropriate equation, as summarized in Table 4.1. Note that the equations in Table 4.1 assume that the flow regimes for the passage of air and water through an integrity breach are the same (i.e., both either laminar or turbulent). If this assumption is not considered appropriate for a specific system and/or application such that inaccurate estimates for direct integrity test sensitivity may result, a hybrid approach may be considered, as described in Appendix C. Other assumptions utilized in the derivation of the ALCR equations in Table 4.1 may also not be valid for some membrane filtration systems. Section C.5 provides some examples of these types of systems and along with general guidance for modifying the ALCR formula derivations to accommodate these systems.

Table 4.1 Approaches for Calculating the ALCR

Module Type	Defect Flow Regime	Model	ALCR Equation	Appendix C Equation
Hollow-fiber[1]	Turbulent[2]	Darcy pipe flow	$170 \bullet Y \bullet \sqrt{\frac{(P_{test} - BP) \bullet (P_{test} + P_{atm})}{(460+T) \bullet TMP}}$	C.4
	Laminar	Hagen-Poiseuille[3]	$\frac{527 \bullet \Delta P_{eff} \bullet (175 - 2.71 \bullet T + 0.0137 \bullet T^2)}{TMP \bullet (460+T)}$	C.15
Flat sheet[4]	Turbulent	Orifice	$170 \bullet Y \bullet \sqrt{\frac{(P_{test} - BP) \bullet (P_{test} + P_{atm})}{(460+T) \bullet TMP}}$	C.9
	Laminar	Hagen-Poiseuille[3]	$\frac{527 \bullet \Delta P_{eff} \bullet (175 - 2.71 \bullet T + 0.0137 \bullet T^2)}{TMP \bullet (460+T)}$	C.15

1 Or hollow-fine-fiber
2 Typically characteristic of larger diameter fibers and higher differential pressures
3 The binomial in the Hagen-Poiseuille equation (C.15) approximates the ratio of water viscosity to air viscosity and is valid for temperatures ranging from approximately 32 to 86 °F. Additional details are provided in Appendix C.
4 Includes spiral-wound and cartridge configurations

The various parameters given in the ALCR equations listed in Table 4.1 include the following:

Y	=	net expansion factor for compressible flow through a pipe to a larger area (dimensionless)
P_{test}	=	direct integrity test pressure (psi)
BP	=	backpressure on the system during the direct integrity test (psi)
P_{atm}	=	atmospheric pressure (psia)
T	=	water temperature (^{o}F)
TMP	=	transmembrane pressure (psi)
ΔP_{eff}	=	effective integrity test pressure (psi)

Additional guidance for both calculating the ALCR and determining appropriate values for the component parameters of the respective ALCR equations (as listed above) is provided in Appendix C, along with derivations of the ALCR equations. Further information regarding the net expansion factor (Y) may be found in various hydraulics references, including Crane (1988). Note that although the Darcy and orifice equations for the ALCR are identical (as shown in Table 4.1), the method for determining the net expansion factor (Y) is different for these two models, as described in Appendix C.

The ALCR can also be determined via empirical means, which would be applicable to any flow regime or configuration of membrane material and are independent of a particular hydraulic model. Some manufacturers may have developed empirical models that could be used to determine the ALCR. If an empirical approach is preferred for determining the ALCR, and a valid empirical model is not available for the system, it may be necessary to develop one. One conceptual procedure for empirically deriving the ALCR for hollow-fiber membrane filtration systems is the correlated airflow measurement (CAM) technique; the details of this procedure are presented in Appendix D.

Measuring the Threshold Direct Integrity Test Response Experimentally

The smallest measurable response of a pressure-based direct integrity test that is associated with a known breach can be evaluated experimentally if this information is not available from the membrane filtration system manufacturer. For pressure-based tests, this response corresponds to the value of ΔP_{test} that should be used in the calculation of sensitivity. This experimental evaluation involves intentionally compromising system integrity in small, discrete, and quantifiable steps and monitoring the corresponding integrity test responses. In the case of microfiltration (MF) and ultrafiltration (UF) systems, several fiber-cutting studies conducted to evaluate the threshold response of various direct integrity tests have been documented in the literature (Adham et al. 1995; Landsness 2001). In general, the procedure for measuring the threshold response experimentally involves the following steps:

1. The membrane system is determined to be integral through the application of a direct integrity test.

2. The investigator intentionally compromises a membrane to generate a known defect. Examples of such compromises include generating a hole in the membrane using a pin of a known diameter or cutting a hollow fiber at a predetermined location. In order to identify the threshold response, it is desirable to utilize a small integrity breach, such as a single cut fiber in a membrane unit.

3. After compromising the membrane, the integrity of the membrane unit is measured using the designated direct integrity test.

4. The process is repeated with additional defects of progressively increasing size or quantity until a measurable response from the direct integrity test is detected. This minimum measurable response represents ΔP_{test} for the purposes of calculating the sensitivity of a direct integrity test.

4.3.1.3 Diffusive Losses and Baseline Decay

If it is determined to be appropriate for the membrane under consideration, the diffusive losses that constitute the baseline integrity test response may be subtracted from the smallest measurable response associated with an integrity breach for the purpose of determining sensitivity. For example, if a pressure decay rate of 0.05 psi/min is typical for an integral membrane unit and the limitations of the test are such that the smallest pressure decay rate that can be reliably associated with an integrity breach is 0.12 psi/min, the incremental response associated with an integrity breach is 0.07 psi/min, and this value may be used in the sensitivity calculation, as illustrated in Equation 4.10 (a variation of Equation 4.8).

$$Q_{air} = \frac{(\Delta P_{test} - D_{base}) \bullet V_{sys}}{P_{atm}} \qquad \text{Equation 4.10}$$

Where:

Q_{air}	=	flow of air (L/min)
ΔP_{test}	=	rate of pressure decay during the integrity test (psi/min)
D_{base}	=	baseline pressure decay (psi/min)
V_{sys}	=	volume of pressurized air in the system during the test (L)
P_{atm}	=	atmospheric pressure (psia)

In general, diffusive losses are most likely to be observed over the duration of the direct integrity test for thin-skinned, asymmetric membranes. The manufacturer should be able to provide information regarding whether or not diffusive losses are expected to be significant. If a high level of sensitivity is required for a membrane filtration system, baseline diffusive losses may be an important consideration. If diffusive losses are determined to be significant and the membrane manufacturer cannot provide a value for the baseline decay, this section describes a means to estimate that decay.

For porous MF, UF, and membrane cartridge filtration (MCF) membranes, diffusive losses occur during direct integrity testing because a certain amount of compressed air used during the test dissolves into the water in the fully-wetted pores and is transported across the membrane surface. In order to calculate the diffusive losses, it can be assumed that the water fills the pores of the membrane and forms a film of thickness z, and that diffusion directly through the membrane material itself is insignificant in comparison to the diffusion across the film of water. Using these assumptions, Equation 4.11 illustrates the relationship between diffusive losses and the various parameters that influence these losses.

$$Q_{diff} = 6 \bullet \left(\frac{D_{aw} \bullet A_m \bullet (P_{test} - BP) \bullet H \bullet \varepsilon}{z} \right) \bullet \left(\frac{R_{gas} \bullet T}{P_{atm}} \right) \qquad \text{Equation 4.11}$$

Where:	Q_{diff}	=	diffusive flow of air through the water held in the membrane pores (L/min)
	D_{aw}	=	diffusion coefficient for air in water (cm^2/s)
	A_m	=	total membrane surface area to which the direct integrity test is applied (m^2)
	P_{test}	=	membrane test pressure (psi)
	BP	=	backpressure on the system during the test (psi)
	H	=	Henry's constant for air-water system (mol/psi-m^3)
	ϵ	=	membrane porosity (dimensionless decimal)
	z	=	membrane thickness (mm)
	R_{gas}	=	universal gas constant (L-psia/mol-K)
	T	=	water temperature during direct integrity test (K)
	P_{atm}	=	atmospheric pressure (psia)
	6	=	unit conversion factor

Note that the dimensionless porosity (ϵ) is defined as the ratio of area of pores to the total membrane area in the unit. This term should not be confused with the pore size of porous MF, UF, and MCF membranes, which is given in terms of the limiting dimension of the openings in the membrane. The porosity of the membrane material can typically be provided by the manufacturer, if necessary. In addition, the diffusion flow path, which is affected by porosity, tortuosity, and the differential pressure across the membrane, is approximated by the membrane thickness (z) in Equation 4.11. Because membrane porosity and tortuosity may be difficult to measure, thus making it problematic to accurately quantify the actual length of the diffusion flow path, a more precise empirical method accounting for these two factors as a combined term has been developed by Farahbakhsh (2003). Both the diffusion coefficient (D_{aw}) and Henry's constant (H) vary with temperature, and Henry's constant also varies somewhat with the concentration of dissolved solids in the water. However, these affects may partially offset and may not be significant. Values for D_{aw} and H as a function of these variables (as applicable) may be found in standard tables in the literature.

The parameters given above for Equation 4.11 are applicable to flat sheet porous membranes, such as those used in membrane cartridge configurations. For porous membranes in a hollow-fiber configuration, such as most MF and UF systems, the following modifications are required:

A_m = log mean total membrane area to which the direct integrity test is applied (m^2)

= $(A_2 - A_1)/\ln(A_2/A_1)$

A_1 = total membrane surface area to which the direct integrity test is applied based on the ***inside*** fiber diameter

A_2 = total membrane surface area to which the direct integrity test is applied based on the ***outside*** fiber diameter

z = differential fiber radius (mm)

= $r_2 - r_1$

r_1 = inside radius of the hollow fiber

r_2 = outside radius of the hollow fiber

It is generally accepted that diffusion of air across the membrane can reduce the measured LRV_{DIT}. Under most circumstances, the amount of diffusion is small in comparison to the flow of air through a breach. However, if the membrane has a propensity to diffuse a significant amount of air (e.g., if the porosity is unusually high) or if a high level of sensitivity is required, it may be necessary to account for diffusive losses. Typically, there is only limited information available regarding the amount of diffusion that occurs for membrane processes used in water treatment under production conditions; however, MF/UF membrane manufacturers can typically provide a value for the baseline decay during a pressure-based direct integrity test for their specific proprietary systems, so it is generally not necessary to explicitly calculate diffusive losses using Equation 4.11.

In the absence of information provided by the manufacturer, it may be advantageous to directly measure the baseline pressure decay on an integral membrane unit, a process that should be conducted using clean membranes to avoid the potential for fouling to artificially hinder diffusion. Because the diffusive loss is directly proportional to temperature (as shown in Equation 4.11), and the diffusion coefficient (which is also directly proportional to diffusive loss) also increases with temperature, it is also important to characterize the baseline decay for the membrane filtration system at a typical water temperature in order to generate an appropriately representative value for diffusive loss. This temperature of evaluation should be recorded for future reference.

With semi-permeable nanofiltration (NF) and reverse osmosis (RO) membranes, diffusive losses occur via the diffusion of air through the saturated membrane material itself. However, because NF and RO modules are manufactured separately from the accompanying filtration systems and inserted manually into pressure vessels, small seal leaks may occur that can be difficult to distinguish from baseline decay. Thus, it is recommended that baseline response for a pressure-based direct integrity test be evaluated for each unit in a NF/RO system on a site-specific basis.

If the membrane module manufacturer has information for typical diffusion airflow rates per unit of membrane area, the expected diffusive airflow for the entire membrane unit should be calculated and compared against the baseline observed during the unit-specific evaluation. If the observed baseline is significantly higher than the predicted diffusive losses, the result could be indicative of an integrity problem, and diagnostic testing (see section 4.8) may be necessary to identify the source of additional airflow or pressure loss. If the membrane module manufacturer is only able to provide a diffusion coefficient for air through the membrane material, Equation 4.12 may be used to estimate the diffusive airflow for a membrane unit with a known membrane area, if necessary:

$$Q_{diff} = 6 \bullet \left(\frac{D_{am} \bullet A_m \bullet (P_{test} - BP) \bullet H}{z} \right) \bullet \left(\frac{R_{gas} \bullet T}{P_{atm}} \right) \qquad \text{Equation 4.12}$$

Where:

Q_{diff}	=	diffusive flow of air through a saturated semi-permeable membrane (L/min)
D_{am}	=	diffusion coefficient for air through a saturated semi-permeable membrane material (cm^2/s)
A_m	=	total membrane surface area to which the direct integrity test is applied (m^2)
P_{test}	=	membrane test pressure (psi)
BP	=	backpressure on the system during the test (psi)
H	=	Henry's constant for air-water system (mol/psi-m^3)
z	=	membrane thickness (mm)
R_{gas}	=	universal gas constant (L-psia/mol-K)
T	=	water temperature during direct integrity test (K)
P_{atm}	=	atmospheric pressure (psia)
6	=	unit conversion factor

Note that the equations for the diffusion of air through porous and semi-permeable membranes – Equations 4.11 and 4.12, respectively – are very similar. These equations differ only in that Equation 4.12 for semi-permeable membranes does not require the membrane porosity and utilizes a diffusion coefficient for a composite membrane layer consisting of both the membrane and the water of saturation bound in the microscopic interstices of the membrane material.

Because spiral-wound NF/RO membranes are typically composite structures consisting of two or more layers (as described in section 2.3.1), it is important that the membrane thickness (z) used corresponds to the layer to which the diffusion coefficient (D_{am}) provided by the membrane manufacturer applies. For example, the diffusion coefficient may apply to the thin, active, semi-permeable layer, in which case the membrane thickness would correspond to this layer only. Alternatively, if the diffusion coefficient is a composite representing the diffusion of air through all the layers of the membrane taken as whole, then the thickness used should be that of the entire membrane, including all layers.

Both Equations 4.11 and 4.12 show that diffusive airflow is directly proportional to membrane area, the applied direct integrity test pressure, and the system backpressure. As a result, the decay should be quantified for each membrane unit of different size in the system and also recalculated if either the applied test pressure or system backpressure are modified.

If the sensitivity is calculated using the ALCR approach described in section 4.3.1.2 and if diffusion is significant, Equation 4.6 can be modified to compensate for diffusive airflow as shown in Equation 4.13:

$$Q_{breach} = \frac{Q_{air} - Q_{diff}}{ALCR} \qquad \text{Equation 4.13}$$

Where:

Q_{breach}	=	flow of water through the critical breach during filtration (L/min)
Q_{air}	=	flow of air through the critical breach during a pressure-based direct integrity test (L/min)
Q_{diff}	=	diffusive flow of air (L/min)
ALCR	=	air-liquid conversion ratio (dimensionless)

Note that the flow of air through the critical breach during a pressure-based direct integrity test (Q_{air}) includes diffusive losses and thus should always be larger than the diffusive flow of air (Q_{diff}). Equation 4.13 should not be used if the baseline decay has already been accounted for by subtracting this baseline decay from the total rate of pressure decay (ΔP_{test}) observed during the integrity test using Equation 4.10, as described at the beginning of this section.

Combining Equation 4.13 with Equation 4.7 yields Equation 4.14, which enables the calculation of sensitivity using the ALCR approach, taking into account diffusive losses.

$$LRV_{DIT} = \log\left(\frac{Q_p \bullet ALCR}{VCF \bullet (Q_{air} - Q_{diff})}\right)$$ Equation 4.14

Where: LRV_{DIT} = direct integrity test sensitivity in terms of LRV (dimensionless)
Q_p = filtrate flow (L/min)
ALCR = air-liquid conversion ratio (dimensionless)
VCF = volumetric concentration factor (dimensionless)
Q_{air} = flow of air (L/min)
Q_{diff} = diffusive flow of air (L/min)

Note that Equations 4.13 and 4.14 are applicable to MCF, MF/UF, and NF/RO membrane filtration systems.

4.3.2 Marker-Based Tests

Sensitivity for marker-based direct integrity tests is determined via a straightforward calculation of the log removal value, similar to the determination of log removal values during a challenge study, as shown in Equation 4.15 (also Equation 3.7):

$$LRV_{DIT} = \log(C_f) - \log(C_p)$$ Equation 4.15 (3.7)

Where: LRV_{DIT} = direct integrity test sensitivity in terms of LRV (dimensionless)
C_f = feed concentration (number or mass / volume)
C_p = filtrate concentration (number or mass / volume)

In using Equation 4.15 to calculate the sensitivity of a marker-based test, the LT2ESWTR specifies that the feed concentration, C_f, is the typical feed concentration of the marker used in the test, and that the filtrate concentration, C_p, is the baseline filtrate concentration of the marker from an integral membrane unit. If the marker is not detected in the filtrate, the term C_p must be set equal to the detection limit. Unlike the sensitivity calculations for pressure-based tests, Equation 4.15 does not incorporate a VCF. This term is not necessary to account for concentration effects on the feed side of the membrane in association with marker-based tests because the concentration of the marker in the filtrate is measured. This measurement would directly account for any increase in the quantity of the marker passing through an integrity breach as a result of feed side concentration effects.

Due to variability in dosing the marker, the day-to-day LRV is likely to vary during system operation. Thus, in order to establish the sensitivity of a marker-based direct integrity test using Equation 4.15, it is necessary to assume an appropriately conservative feed concentration such that the LRVs determined on a day-to-day basis meet or exceed the LRV_{DIT} unless there is an integrity breach. An example of such a conservative approach would be the use of the lower bound of the anticipated concentration range of seeded marker as the feed concentration for the purposes of calculating the sensitivity of the test. Note that a marker-based direct integrity test must utilize a feed concentration sufficient to demonstrate the required *Cryptosporidium* LRV for rule compliance. Because a typical feed water will normally not contain a sufficient number of particles at the required 3-μm resolution or else will not be sufficiently characterized to demonstrate this, seeding will usually be required for marker-based tests.

In order to optimize the sensitivity and reliability of a marker-based direct integrity test, it is important to use an accurate method for quantifying the feed and filtrate concentrations. Since the feed and filtrate concentrations will differ by orders of magnitude, an analytical method with a wide dynamic range is desired. If such a range is not available using a single device, two different instruments may need to be used to measure these respective concentrations. Regardless of the dynamic range of the instrument(s), it is likely that different analytical volumes will need to be used to deal with the different concentration ranges; however, the concentrations will have to be expressed in terms of equivalent volumes for the purpose of calculating an LRV. Some specific considerations regarding the use of particulate and molecular marker-based direct integrity tests are discussed in section 4.7.5.

4.4 Test Frequency

Most currently available direct integrity tests require the membrane unit to be taken off-line for testing and thus are conducted in a periodic manner, requiring a balance between the need to verify system integrity with the desire to minimize system downtime and lost productivity. In addition, although some marker-based tests may be conducted while the membrane unit is on-line and in production, it is generally neither practical nor cost effective to implement these tests on a continual basis. Thus, the frequency at which direct integrity testing is conducted for membrane filtration systems represents a compromise between these competing objectives.

The LT2ESWTR requires that direct integrity testing be conducted on each membrane unit at least once each day that the membrane unit is in operation for rule compliance, unless the State approves less frequent testing (40 CFR 141.719(b)(3)(vi)). This minimum test frequency is intended to balance the need to verify system integrity as often as possible with cost and production considerations and is based in part on a USEPA report that indicated daily direct integrity testing was relatively common practice at membrane filtration facilities (USEPA 2001). It is important to note that the rule requires that each unit be subjected to direct integrity testing on a daily basis, even if the unit is operational for only a fraction of each day. It is also recommended that a given unit be tested at approximately the same time each day (site-specific facility operations permitting) in order to maintain a roughly consistent time interval between applications of direct integrity testing; however, this is not specifically required by the rule.

The State may require either more or less frequent integrity testing for compliance with the rule at its discretion, although less frequent testing must be supported by demonstrated process reliability, the use of multiple barriers effective for *Cryptosporidium*, or reliable process safeguards (40 CFR 141.719(b)(3)(vi)). For example, in terms of process reliability, the State may opt to reduce the frequency of direct integrity testing if the membrane filtration system has a significant history of demonstrated operation either without the detection of an integrity breach or with only rare occurrences of integrity breaches that have never been large enough to compromise the ability of the system to achieve the awarded removal credit. Alternatively, the State may also reduce the required test frequency if the overall treatment scheme (including both membrane filtration and other processes) incorporates at least one additional process that is capable of achieving a substantial portion of the required *Cryptosporidium* treatment credit. In this case, even with a small integrity breach the State may be confident that the multiple barrier treatment process is fully capable of achieving compliance with the LT2ESWTR.

The State may also permit a reduced direct integrity test frequency if other safety factors are utilized to mitigate the risk associated with a potential integrity breach. One such strategy might be maintaining filtrate storage with a detention time equivalent to or longer than the time between direct integrity test events. If an integrity breach were detected, this storage would provide the utility with sufficient time to take the necessary mitigation measures to ensure that any contamination risks are addressed before the water enters the distribution system. Another possibility might be the use of continuous indirect integrity monitoring techniques with very high demonstrated sensitivities relative to other indirect monitoring methods. This strategy would represent a trade-off in that although the more sensitive (relative to the indirect monitoring methods) direct integrity test would be conducted less often, the ability to detect an integrity breach between applications of the direct test would be increased.

Note that any reduction in direct integrity test frequency is subject to the discretion of the State, which may utilize any of these suggested strategies as the basis for its decision or any other criteria it determines to be appropriate. Also, although unrelated to rule compliance, more frequent testing may be appropriate under certain specific circumstances, such as during initial facility start-up, as described in Chapter 8. More frequent direct integrity testing may also be advantageous for systems that rely on membrane filtration to achieve a high log removal of *Cryptosporidium*, since the risk associated with an integrity breach is greater. Conducting direct integrity testing on a membrane unit more frequently than once per day may be voluntarily implemented by a utility or required at the discretion of the State.

4.5 Establishing Control Limits

A control limit (CL) is defined as a response that, if exceeded, indicates a potential problem with the system and triggers a response. Multiple CLs can be set at different levels to indicate the severity of the potential problem. In the context of direct integrity testing, CLs are set at levels associated with various degrees of integrity loss. Under the provisions of the LT2ESWTR, a CL within the sensitivity limits of the direct integrity test must be established at the threshold test response that is indicative of an integral membrane unit capable of achieving the *Cryptosporidium* removal credit awarded by the State for rule compliance

(40 CFR 141.719(b)(3)(iv)). Because utilities or States would have the option to implement a series of tiered CLs that may represent progressively greater levels of integrity loss leading up to the specific CL required under the rule, in this guidance manual the LT2ESWTR-mandated CL is referred to as an upper control limit (UCL). If the integrity test response is below the UCL, the membrane unit should be achieving a LRV equal to or greater than the removal credit awarded to the process. Alternatively, if the UCL is exceeded, the membrane unit is required to be taken off-line for diagnostic testing (as described in section 4.8) and repair.

The same principles used to establish direct integrity test sensitivity are also used to establish the UCL. For ***pressure-based tests***, the UCL may be calculated using the ALCR methodology. A modified version of Equation 4.7 yields an expression for the UCL for, as shown in Equation 4.16:

$$UCL = \frac{Q_p \bullet ALCR}{10^{LRC} \bullet VCF} \qquad \text{Equation 4.16}$$

Where:

UCL	=	upper control limit in terms of airflow (L/min)
Q_p	=	membrane unit design capacity filtrate flow (L/min)
ALCR	=	air-liquid conversion ratio (dimensionless)
LRC	=	log removal credit awarded (dimensionless)
VCF	=	volumetric concentration factor (dimensionless)

Similarly, Equation 4.9 can be rearranged to establish an expression for calculating the UCL in terms of a pressure decay rate, as shown in Equation 4.17:

$$UCL = \frac{Q_p \bullet ALCR \bullet P_{atm}}{10^{LRC} \bullet V_{sys} \bullet VCF} \qquad \text{Equation 4.17}$$

Where:

UCL	=	upper control limit in terms of pressure decay rate (psi/min)
Q_p	=	membrane unit design capacity filtrate flow (L/min)
ALCR	=	air-liquid conversion ratio (dimensionless)
P_{atm}	=	atmospheric pressure (psia)
LRC	=	log removal credit (dimensionless)
V_{sys}	=	volume of pressurized air in the system during the test (L)
VCF	=	volumetric concentration factor (dimensionless)

Values for the parameters in Equations 4.16 and 4.17 should be the same as the analogous terms used to calculate sensitivity using Equations 4.7 and 4.9, respectively. Note that to the extent possible, these values should be selected to yield a conservative result for the UCL.

Equations 4.16 and 4.17 establish the maximum direct integrity test response that can be used as a UCL for the log removal credit (LRC) awarded by the State. The LRC, in turn, must be less than or equal to the lower value of either the log removal value determined during challenge testing ($LRV_{C\text{-}Test}$) or the sensitivity of the direct integrity test used (LRV_{DIT}).

In the context of Equations 4.16 and 4.17, CLs are expressed in terms of the actual response from the direct integrity test (i.e., flow of air or pressure decay rate, respectively). In this form the CLs may be most useful for operators, since these could be directly compared to integrity test results. However, it may also be useful to calculate the corresponding LRVs for both the CL(s) and the individual direct integrity test responses using the generic forms of Equations 4.7 (for tests yielding results in terms of the flow of air) or 4.9 (for tests yielding results in terms of pressure decay). In this case these equations will simply yield a LRV corresponding to a particular direct integrity test result (i.e., a general LRV) rather than the sensitivity of the test (i.e., LRV_{DIT}). Many membrane systems have automated data acquisition equipment that could be programmed to calculate the LRV based upon the results of the most recent integrity test results and current operating conditions. These parameters may be displayed and trended to track system performance. Additional guidance on data analysis is provided in section 4.9.

For ***marker-based tests***, which yield results in terms of log removal, the UCL is simply equal to the log removal credit awarded by the State.

Any CLs other than that mandated by the LT2ESWTR that are either voluntarily implemented by the utility or required by the State are referred to as lower control limits (LCLs). For example, a LCL may be established to provide operators with an indication that there may be an integrity breach before the breach becomes a compliance concern. In this scenario the LCL could be used in the context of preventative maintenance, and excursions above the LCL could prompt investigation and repair during scheduled downtime for the unit rather than require an immediate shutdown. The use of CLs and integrity testing in the context of a comprehensive integrity verification program is discussed further in Appendix A.

Unlike the UCL that is established by the log removal credit awarded to the process, a LCL can be established based on the needs and objectives of the utility. However, any LCLs should be above the baseline integrity test response for an integral membrane and below the UCL in order to be useful. The baseline integrity test value for a membrane unit can be established during the commissioning of the facility, after the membrane system has been fully wetted and determined to be integral. The baseline level is described as the normal range of direct integrity test results that would occur for an integral membrane unit. A practical lower bound for any LCLs is the sensitivity of the direct integrity test (i.e., the lowest response that can be reliably measured that is indicative of an integrity breach).

As with the determination of membrane unit sensitivity, for systems that utilize multiple stages of membrane filtration, the UCL for each stage must be determined independently. Thus, for pressure-based tests, the parameters used to calculated the UCL (e.g., log removal credit, ALCR, VCF, etc.) must be specific to the membrane units in each respective stage. This requirement is applicable to all stages to which the LT2ESWTR applies (i.e., those that produce filtrate for drinking water or post-filtration treatment rather than for recycling to an upstream point in the treatment process).

Appendix C:
Calculating the Air-Liquid Conversion Ratio

C.1 Introduction

As described in Chapter 4, the regulatory framework for the Long-Term 2 Enhanced Surface Water Treatment Rule (LT2ESWTR) requires that the flow through the smallest integrity breach that generates a measurable response from the direct integrity test (i.e., the critical breach size - Q_{breach}) be determined in order to establish the sensitivity of a pressure-based test method (40 CFR 141.719(b)(3)(iii)(A)). (Under the LT2ESWTR, sensitivity is defined as the maximum log removal value (LRV) that can be reliably verified by the direct integrity test (i.e., LRV_{DIT}).) However, because most pressure-based direct integrity tests yield results in terms of airflow or pressure decay, it may be necessary to convert these results to an equivalent value for the flow of water through the critical breach under typical filtration conditions. This conversion is necessary for calculating both the sensitivity and the upper control limit (UCL) for a pressure-based direct integrity test.

Although there are a number of methods for converting a direct integrity test response to a corresponding flow of water, each can be generally categorized as one of two types of approaches: mathematical modeling or experimental determination. This appendix describes a mathematical approach based on a parameter called the air-liquid conversion ratio (ALCR), which is defined as the ratio of air that would flow through a breach during a direct integrity test to the amount of water that would flow through the breach during filtration, as shown in Equation C.1 (also Equation 4.5):

$$ALCR = \left(\frac{Q_{air}}{Q_{breach}} \right) \qquad \text{Equation C.1 (4.5)}$$

Where:

ALCR	=	air-liquid conversion ratio (dimensionless)
Q_{air}	=	flow of air through the critical breach during a pressure-based direct integrity test (volume / time)
Q_{breach}	=	flow of water through the critical breach during filtration (volume / time)

Because of the many variations in membrane configurations, breaches in the membrane may exhibit either turbulent or laminar flow characteristics depending upon the location and size of the defect, as well as the pressure differential between the feed and filtrate. In addition, there are fundamental differences between hollow-fiber and flat sheet membrane breaches, since the most common breaches associated with hollow-fiber modules exhibit pipe flow characteristics, while flat sheet breaches are best represented by an orifice model. Consequently, three different hydraulic models have been developed for determining the ALCR for a particular membrane system, depending on the configuration of the membrane material (i.e., hollow-fiber vs. flat sheet) and the type of flow (i.e., laminar vs. turbulent) that is expected through the critical breach. These three models include the Darcy pipe flow model (for breaches in a hollow-fiber (or hollow-fine-fiber) module under conditions of turbulent flow), the orifice model (for modules utilizing flat sheet membranes such as spiral-wound and membrane cartridge configurations under conditions of turbulent flow), and the Hagen-Poiseuille model (for any configuration under conditions of laminar flow). Table C.1 summarizes the various approaches for calculating the ALCR based on these three models and the conditions under which the use of each model is appropriate.

Table C.1 Approaches for Calculating the ALCR

Module Type	Defect Flow Regime	Model
Hollow-fiber[1]	Turbulent[2]	Darcy Pipe Flow
	Laminar	Hagen-Poiseuille
Flat sheet[3]	Turbulent	Orifice
	Laminar	Hagen-Poiseuille

1 Or hollow-fine-fiber
2 Typically characteristic of larger diameter fibers and higher differential pressures
3 Includes spiral-wound and cartridge configurations

Note that the various methods presented in this appendix for determining the ALCR implicitly assume that the flow regime for airflow through a breach during direct integrity testing is the same as that for liquid flow though a breach during filtration (i.e., either both laminar or both turbulent). If this assumption is determined to be inappropriate for a given membrane filtration system such that inaccurate and non-conservative estimates for sensitivity may result, then a hybrid approach may be considered. An example of such a hybrid approach is to assume laminar water flow and turbulent airflow, which could be modeled through the application of the Hagen-Poiseuille equation for water and the Darcy equation for air. The ALCR for such cases could then be derived using a similar methodology to that described in sections C.2, C.3, and C.4.

Procedures for calculating the ALCR and subsequently the sensitivity and UCL for applicable pressure-based direct integrity tests are given in Chapter 4, but the derivations of the various hydraulic models that form the basis for the respective ALCR equations are provided in the following sections of this appendix. Note that while the derivation of the ALCR equations relies on various hydraulic models that could be used to directly calculate the flow of air (Q_{air}) and water (Q_{breach}) through an integrity breach, direct application of these equations requires knowledge of the critical breach size, which is difficult and impractical to accurately quantify.

The advantage of the ALCR is that the terms relating to the size of the breach cancel out, yielding equations for the ALCR that are a function of either known and/or more easily determined parameters and independent of the critical breach size or geometry. Thus, although the ALCR equations in this appendix are derived for conditions assuming the flow of air (Q_{air}) and water (Q_{breach}) through the critical breach, the ALCR is independent of integrity breach size (i.e., physical dimensions of the breach) or magnitude (i.e., number of distinct breaches), and thus is a scalable parameter. An additional section is also included at the end of this appendix (C.5) that discusses cases in which the ALCR equations derived in sections C.2, C.3, and C.4 may not be applicable for some membrane filtration systems, as well as modifying the derivations to accommodate these systems.

For the derivation of the various models described in this appendix, airflow equations have been developed using a standard temperature and pressure of 68 °F (528 R, 293 K, or 20 °C) and 0 psi (1 atm or 14.7 psia). These standard conditions were selected to be consistent with the convention for airflow measurement devices. The equations can be modified to a different set of reference conditions through application of the ideal gas law expressed in terms of absolute temperature and pressure, if necessary. In addition, the temperature of the air used in a pressure-based direct integrity test is assumed to be the same as that for the water in the membrane filtration system, since these temperatures are expected to rapidly equilibrate.

Additional background on the hydraulic modeling developed in this appendix may be found in Crane's *Flow of Fluids Through Valves, Fittings, and Pipes* (1988). All of the basic hydraulic equations used in the derivation of the ALCR equations are included in the Crane text.

C.2 Darcy Pipe Flow Model

The Darcy pipe flow model is used to describe turbulent flow through an integrity breach with characteristics similar to a broken hollow-fiber. Generally, turbulent flow may be expected through larger diameter broken fibers and at higher differential pressures. The Darcy equations for the flow of air and water through a pipe are given in Equations C.2 and C.3, respectively:

$$Q_{air} = 11.3 \bullet Y \bullet d_{fiber}^{\ 2} \bullet \sqrt{\frac{(P_{test} - BP) \bullet (P_{test} + P_{atm})}{(460 + T) \bullet K_{air}}} \qquad \text{Equation C.2}$$

Where:

Q_{air}	=	flow of air at standard conditions (ft^3/s)
Y	=	net expansion factor for compressible flow through a pipe to a larger area (dimensionless)
d_{fiber}	=	fiber diameter (in)
P_{test}	=	integrity test pressure (psi)
BP	=	backpressure on the system during the integrity test (psi)
P_{atm}	=	atmospheric pressure (psia)
T	=	water temperature (°F)
K_{air}	=	resistance coefficient of air (dimensionless)

$$Q_{breach} = 0.525 \bullet d_{fiber}^{\ 2} \bullet \sqrt{\frac{TMP}{K_{water} \bullet \rho_w}}$$ Equation C.3

Where:

Q_{breach}	=	flow of water through the critical breach during filtration (ft^3/s)
d_{fiber}	=	fiber diameter (in)
TMP	=	transmembrane pressure (psi)
K_{water}	=	resistance coefficient of water (dimensionless)
ρ_w	=	density of water (lbs/ft^3)

Assuming that the resistance coefficients for air and water are similar (i.e., $K_{air} \approx K_{water}$) and applying a value of 62.4 lbs/ft^3 for the density of water, the ratio of Equation C.2 to Equation C.3 yields an expression for the ALCR, as given by Equation C.4:

$$ALCR = 170 \bullet Y \bullet \sqrt{\frac{(P_{test} - BP) \bullet (P_{test} + P_{atm})}{(460 + T) \bullet TMP}}$$ Equation C.4

Where:

ALCR	=	air-liquid conversion ratio (dimensionless)
Y	=	net expansion factor for compressible flow through a pipe to a larger area (dimensionless)
P_{test}	=	direct integrity test pressure (psi)
BP	=	backpressure on the system during the integrity test (psi)
P_{atm}	=	atmospheric pressure (psia)
T	=	water temperature (^oF)
TMP	=	transmembrane pressure during normal operation (psi)

The ALCR is used in the equations for determining both the sensitivity and the UCL for pressure-based direct integrity tests, as described in Chapter 4. Consequently, the values of the parameters in Equation C.4 should be selected to yield a lower, more conservative value for the ALCR. For example, the transmembrane pressure (TMP) in Equation C.4 comes from the expression for Q_{breach} (Equation C.3) during filtration, and thus the most conservative ALCR result would be generated from using the maximum anticipated TMP during normal operation.

The net expansion factor for compressible flow (Y) may be obtained from charts in various hydraulics references, such as Crane (1988) (page A-22). Using the appropriate chart for airflow, values for Y are given as a function of pressure and the flow resistance coefficient, as shown in Equation C.5 (a non-specific expression illustrating the relationship between Y and its variables):

$$Y \propto \left[\frac{1}{\left(\frac{P_{test} - BP}{P_{test} + P_{atm}} \right)}, K \right]$$ Equation C.5

Where:

Y	=	net expansion factor for compressible flow through a pipe to a larger area (dimensionless)
P_{test}	=	direct integrity test pressure (psi)
BP	=	backpressure on the system during the integrity test (psi)
P_{atm}	=	atmospheric pressure (psia)
K	=	flow resistance coefficient (dimensionless)

The flow resistance coefficient (K) is a common fluid flow parameter described by most hydraulics texts and is defined as shown in Equation C.6:

$$K = f \bullet \frac{L}{d_{fiber}}$$ Equation C.6

Where:

K	=	flow resistance coefficient (dimensionless)
f	=	friction factor (dimensionless)
L	=	length of the defect (in)
d_{fiber}	=	fiber diameter (in)

Using the conservative scenario of a fiber break at the point where the fiber enters the pot, the length of the defect (L) is represented by the length of the lumen encasement into the membrane pot. The friction factor (*f*) may be estimated from a Moody diagram or the corresponding tabulated values, both of which are readily in available in most hydraulics references. The relative roughness (e/d_{fiber}) that is required to estimate the value for the friction faction may be calculated by either obtaining a product-specific value for the specific roughness (*e*) from the manufacturer or by using the membrane pore size as an estimate of the specific roughness.

Note that the net expansion factor (Y) should remain constant over time for practical purposes if appropriately conservative values are used to calculate this parameter. Thus, the determination of Y should represent a one-time, site-specific calculation. Also, because the ALCR is directly proportional to Y (as shown in Equation C.4), lower values for Y result in lower, more conservative values for the ALCR.

An iterative solution may be required to determine a value for the net expansion factor (Y). A general outline for one such iterative process is given as follows. The use of a spreadsheet may help facilitate the various calculations required.

1. Select a reasonable value for the friction factor (*f*).

2. Calculate the flow resistance coefficient (K) using Equation C.6.

3. Obtain a value for the Reynolds number (Re) from tabulated values for the friction factor (*f*) as a function of the Reynolds number (Re) and the relative roughness (e/d_{fiber}).

4. Calculate airflow (Q_{air}) from the equation for the Reynolds number (Re) as a function of equivalent diameter, air velocity, and dynamic viscosity (as referenced in fluid mechanics and fluid dynamics texts). For the purposes of determining the equivalent diameter and velocity (i.e., the flow (Q_{air}) divided by the cross-sectional area), assume that the applicable integrity breach may be represented by a pipe (e.g., a hollow fiber) flowing full with air. Use the maximum anticipated temperature and the minimum pressure applied over the duration of the direct integrity test (i.e., accounting for baseline decay) to generate a conservative (i.e., low) value for the dynamic viscosity and thus, in turn a conservative (i.e., low) value for both airflow (Q_{air}) and the ALCR.

5. Using tables available in hydraulics texts (e.g., page A-22 of Crane (1988)), apply the flow resistance coefficient (K) and the pressure ratio to determine a value for the net expansion factor (Y), as shown in Equation C.5.

6. Calculate airflow (Q_{air}) using Equation C.2. Assume that $K \approx K_{air}$.

7. If the airflow (Q_{air}) calculated in steps 4 and 6 (above) is approximately the same, then the net expansion factor (Y) determined in step 5 is correct. Otherwise, select a revised value for the friction factor (*f*) and repeat steps 1-7 in an iterative process until the two calculated values for airflow (Q_{air}) converge.

C.3 Orifice Model

The orifice flow model may be used to approximate turbulent flow through an integrity breach with characteristics similar to a hole in a flat sheet membrane that may be configured as a cartridge or a spiral-wound module. The representative equations for airflow and water flow through an orifice are given as Equations C.7 and C.8, respectively:

$$Q_{air} = 11.3 \bullet Y \bullet d_{fiber}^{\ 2} \bullet C \bullet \sqrt{\frac{(P_{test} - BP) \bullet (P_{test} + P_{atm})}{460 + T}} \qquad \text{Equation C.7}$$

Where:

Q_{air}	=	flow of air at standard conditions (ft^3/s)
Y	=	net expansion factor for compressible flow through a pipe to a larger area (dimensionless)
d_{fiber}	=	fiber diameter (in)
C	=	coefficient of discharge (dimensionless)
P_{test}	=	direct integrity test pressure (psi)
BP	=	backpressure on the system during the integrity test (psi)
P_{atm}	=	atmospheric pressure (psia)
T	=	water temperature (^oF)

$$Q_{breach} = 0.525 \bullet d_{fiber}^{\ 2} \bullet C \bullet \sqrt{\frac{TMP}{\rho_w}} \qquad \text{Equation C.8}$$

Where:

Q_{breach}	=	flow of water through the critical breach during filtration (ft^3/s)
d_{fiber}	=	fiber diameter (in)
C	=	coefficient of discharge (dimensionless)
TMP	=	transmembrane pressure (psi)
ρ_w	=	density of water (lbs/ft^3)

The ratio of Equation C.7 to Equation C.8 yields an expression for the ALCR, as given in Equation C.9. Note that this equation incorporates a value of 62.4 lbs/ft^3 for density of water.

$$ALCR = 170 \bullet Y \bullet \sqrt{\frac{(P_{test} - BP) \bullet (P_{test} + P_{atm})}{(460 + T) \bullet TMP}} \qquad \text{Equation C.9}$$

Where:

ALCR	=	air-liquid conversion ratio (dimensionless)
Y	=	net expansion factor for compressible flow through a pipe to a larger area (dimensionless)
P_{test}	=	direct integrity test pressure (psi)
BP	=	backpressure on the system during the integrity test (psi)
P_{atm}	=	atmospheric pressure (psia)
T	=	water temperature (^oF)
TMP	=	transmembrane pressure during normal operation (psi)

Note that although derivations are slightly different, the resulting ALCR equations for the Darcy and orifice models (Equations C.4 and C.9, respectively) are identical. However, the two models utilize different methodologies for determining the net expansion factor for compressible flow (Y). As described in section C.2 for the Darcy model, the values of the parameters in Equation C.9 should be selected to yield a conservative value for the ALCR. For example, the TMP in Equation C.9 comes from the expression for Q_{breach} (Equation C.8) during filtration, and thus the most conservative ALCR result would be generated from using the maximum anticipated TMP during normal operation.

As with the Darcy model, the net expansion factor for compressible flow (Y) may be obtained from charts in various hydraulics references, such as Crane (1988) (page A-21). However, for the orifice model, Equation C.10 may also be used to calculate the net expansion factor, as follows:

$$Y = 1 - \left[0.293 \bullet \left(1 - \frac{BP + P_{atm}}{P_{test} + P_{atm}} \right) \right] \qquad \text{Equation C.10}$$

Where:

Where:	Y	=	net expansion factor for compressible flow through a pipe to a larger area (dimensionless)
	BP	=	backpressure on the system during the integrity test (psi)
	P_{atm}	=	atmospheric pressure (psia)
	P_{test}	=	direct integrity test pressure (psi)

Because the ALCR is directly proportional to Y (as shown in Equation C.9), lower values for Y result in lower, more conservative values for the ALCR. Also, for practical purposes the net expansion factor (Y) should remain constant over time if appropriately conservative values are used to calculate this parameter. Thus, the determination of Y should represent a one-time, site-specific calculation.

C.4 Hagen-Poiseuille Model

The Hagen-Poiseuille model is appropriate for small integrity breaches (such as a pin hole or a broken, small-diameter hollow-fiber under low differential pressure) that would result in laminar flow. Using this model, the equation for airflow through a small defect under laminar flow conditions is given by Equation C.11:

$$Q_{air} = \frac{49.5 \bullet \pi \bullet d_{fect}^{\ 4} \bullet \Delta P_{eff} \bullet g}{L \bullet \mu_{air} \bullet (460 + T)} \qquad \text{Equation C.11}$$

Where:

Q_{air}	=	flow of air at standard conditions (ft^3/s)
d_{fect}	=	defect diameter (in)
ΔP_{eff}	=	effective integrity test pressure (psi)
g	=	gravitational constant (32.2 lbm-ft/lbf-s^2)
L	=	length of the defect (in)
μ_{air}	=	viscosity of air (lbs/ft-s)
T	=	water temperature (^oF)

Because air is a compressible fluid, the airflow is determined using the effective integrity test pressure, ΔP_{eff}, calculated according to Equation C.12.

$$\Delta P_{eff} = \left[(P_{test} - BP)\right] \bullet \left[\frac{(P_{test} + P_{atm}) + (BP + P_{atm})}{2 \bullet (BP + P_{atm})}\right] \bullet \left[\frac{(BP + P_{atm})}{P_{atm}}\right] \qquad \text{Equation C.12}$$

Where:

ΔP_{eff}	=	effective air integrity test pressure (psi)
P_{test}	=	direct integrity test pressure (psi)
BP	=	backpressure on the system during the integrity test (psi)
P_{atm}	=	atmospheric pressure (psia)

The elements of the effective integrity test pressure include three primary terms, as individually bracketed in Equation C.12. These three terms are, respectively:

- the differential pressure across the membrane during the integrity test
- a term that accounts for the average velocity gradient of the compressed air as it passes across the membrane
- a multiplier that is necessary to convert the backpressure as it leaves the membrane to standard atmospheric conditions

The Hagan-Poiseuille equation for liquid flow through a breach under conditions of laminar flow is shown as Equation C.13:

$$Q_{breach} = \frac{0.094 \bullet \pi \bullet d_{fect}^{\ 4} \bullet g \bullet TMP}{L \bullet \mu_w} \qquad \text{Equation C.13}$$

Where:			
	Q_{breach}	=	flow of water through the critical breach during filtration (ft^3/s)
	d_{fect}	=	defect diameter (in)
	g	=	gravitational constant (32.2 lbm-ft/lbf-s^2)
	TMP	=	transmembrane pressure (psi)
	L	=	length of the defect (in)
	μ_w	=	viscosity of water (lbs/ft-s)

The ratio of Equation C.11 to Equation C.13 yields an expression for the ALCR, as given by Equation C.14:

$$ALCR = \frac{527 \bullet \Delta P_{eff} \bullet \mu_w}{TMP \bullet \mu_{air} \bullet (460 + T)} \qquad \text{Equation C.14}$$

Where:			
	ALCR	=	air-liquid conversion ratio (dimensionless)
	ΔP_{eff}	=	effective integrity test pressure (psi)
	μ_w	=	viscosity of water (lbs/ft-s)
	TMP	=	transmembrane pressure during normal operation (psi)
	μ_{air}	=	viscosity of air (lbs/ft-s)
	T	=	water temperature (^{o}F)

The ratio of the viscosity of water to the viscosity of air (μ_w / μ_{air}) may be combined and expressed as a single function of the water temperature that is derived by fitting a curve to discrete data points for the viscosity ratio. Equation C.15 simplifies the calculation of the ALCR to an expression that is function of only measured pressures and the water temperature. Note that this form of the equation is only valid in the temperature range from 32 to 86 ^{o}F, in accordance with the limitations of the binomial fit for the viscosity ratio. If the temperature is outside of this range, the more general expression in Equation C.14 should be used.

$$ALCR = \frac{527 \bullet \Delta P_{eff} \bullet (175 - 2.71 \bullet T + 0.0137 \bullet T^2)}{TMP \bullet (460 + T)} \qquad \text{Equation C.15}$$

Where:			
	ALCR	=	air-liquid conversion ratio (dimensionless)
	ΔP_{eff}	=	effective integrity test pressure (psi)
	T	=	water temperature (^{o}F)
	TMP	=	transmembrane pressure (psi)

As with the Darcy and orifice models, the values for the parameters ΔP_{eff} and TMP used in Equation C.14 or C.15 should be selected to yield a lower, more conservative value for the ALCR. For example, the TMP in Equation C.15 comes from the expression for Q_{breach} (Equation C.13) during filtration, and thus the most conservative ALCR result would be generated from using the maximum anticipated TMP during normal operation.

C.5 Applicability of ALCR Equations

The various expressions for the ALCR developed in this appendix are each derived under the assumption that water and air pass through the same integrity breach during normal operation and pressure-based direct integrity testing, respectively. This consistency allows the terms relating to breach characteristics to cancel out in the ALCR derivations, resulting in equations that are independent of any specific knowledge of the integrity breach. Although this assumption is consistent with the operation of most membrane filtration systems, there may be some systems that operate in a manner that renders this assumption invalid. In these cases, the ALCR equations described in this appendix are not directly applicable, and the derivation of the ALCR must be modified to match the specific operation of a particular membrane filtration system.

One example of such a scenario might be a hollow-fiber membrane filtration system utilizing modules operating in an inside-out mode in which feed water flows into the fiber lumen from both ends. In this case, a single broken fiber represents two different pathways for water to flow through an integrity breach. If the Darcy model is applicable (i.e., turbulent flow conditions prevail), careful consideration must be given to the selection of the appropriate value for the net expansion factor for compressible flow (Y). This parameter is a function of the defect length (L), which will be different for each of the two flow pathways created by a single broken fiber (assuming the fiber is not severed precisely in the middle). In order to generate the most conservative (i.e., lowest) value for the ALCR (thus resulting in the most conservative direct integrity test sensitivity), the smallest potential value for L (i.e., the shortest length) should be used, which is generally represented by the length of the fiber encasement into the potting material. This value for L should always be utilized to yield a conservative ALCR, independent of the magnitude of the integrity breach. This approximation is reasonable since the flow through the short fiber length (i.e., the length of fiber embedded in the pot) will be substantially greater than the flow through the longer fiber length in most cases. If there is any question as to whether or not this approximation is appropriate for a particular system, the ALCR for different lengths could be calculated to evaluate the sensitivity of this important parameter to the length of the flow path.

A more complex scenario might involve a system similar to one described in the previous example, but which applies pressurized air from only one end of the fiber during direct integrity testing. In this case, a single broken fiber would represent two separate pathways for water to flow through an integrity breach during normal operation, but only a single path for air during integrity testing. This difference must be accounted for in the derivation of an expression for the ALCR. As shown in Equation C.1, the ALCR is the ratio of the air flowing through an integrity breach during direct integrity testing (Q_{air}) to the flow of water through the same breach during normal operation (Q_{breach}). For the system described in this example, any number of broken fibers would generate double that number of pathways for feed water to bypass the membrane filtration process via integrity breaches. Thus, the term Q_{breach} must also be doubled in the ALCR expression for both the Darcy (i.e., turbulent flow) and the Hagen-Poiseuille (i.e., laminar flow) models.

The two examples addressed in this section are just two possible scenarios in which the ALCR equations developed in this appendix may not be directly applicable to some membrane filtration systems. Because this guidance manual cannot anticipate and address every such potential case, it is recommended that each membrane filtration system be evaluated on a site-specific basis to determine whether the given ALCR equations may be used or if the derivation(s) must be modified to accommodate system-specific characteristics. Thus, any deviations from the assumptions used to derive the ALCR equations presented in this document may require a more complex treatment in which the flows of water and air through an integrity breach are determined separately using system-specific assumptions that are valid for each of these two respective flows.

Appendix D:

Empirical Method for Determining the Air-Liquid Conversion Ratio for a Hollow-Fiber Membrane Filtration System

D.1 Introduction

As described in Chapter 4, the regulatory framework for the Long-Term 2 Enhanced Surface Water Treatment Rule (LT2ESWTR) requires that the flow through the smallest integrity breach that generates a measurable response from the direct integrity test (i.e., the critical breach size – Q_{breach}) be determined in order to establish the sensitivity of a pressure-based test method (40 CFR 141.719(b)(3)(iii)(A)). (Under the LT2ESWTR, sensitivity is defined as the maximum log removal value (LRV) that can be reliably verified by the direct integrity test (i.e., LRV_{DIT}).) However, because most pressure-based direct integrity tests yield results in terms of airflow or pressure decay, for systems that utilize such tests it may be necessary to convert these results to an equivalent value for the flow of water through the critical breach under typical filtration conditions. Although there are a number of potential methods for converting a direct integrity test response to a corresponding flow of water, each can be generally categorized as one of two types of approaches: mathematical modeling or experimental determination. This appendix describes an empirical approach based on a parameter called the air-liquid conversion ratio (ALCR), which is defined as the ratio of air that would flow through a breach during a direct integrity test to the amount of water that would flow through the breach during filtration, as shown in Equation D.1:

$$ALCR = \left(\frac{Q_{air}}{Q_{water}}\right) \qquad \text{Equation D.1}$$

Where:

ALCR	=	air-liquid conversion ratio (dimensionless)
Q_{air}	=	flow of air through an integrity breach during a pressure-based direct integrity test (volume / time)
Q_{water}	=	flow of water through an integrity breach during filtration (volume / time)

While Appendix C describes the hydraulic models that could be used to calculate the ALCR for various types of membrane filtration systems and under different flow regimes, this appendix provides an example of an empirical method based on bubble point theory that can be used to determine the ALCR for a system using microporous hollow-fiber membranes – the correlated airflow measurement (CAM) technique. The CAM technique measures the flow of air (Q_{air}) and water (Q_{water}) through a fiber break scenario to empirically determine the ALCR of a membrane filtration unit for a system utilizing a pressure-based direct integrity test. This method is specific to hollow-fiber membrane processes in which the geometry of the fiber and associated module is known.

The CAM methodology involves first measuring the flow of water through a known integrity breach at various transmembrane pressures (TMPs) representative of normal operation, and subsequently measuring the flow of air through the same integrity breach during a pressure-based direct integrity test using a variety of potential test pressures. The data obtained from these measurements may be fitted with respective equations to establish empirical relationships between the applied pressures and the flow of water or air through an integrity breach. These functions can be used to determine the ALCR for any given operating transmembrane pressure (TMP) and direct integrity test pressure (P_{test}) (accounting for backpressure during the test, as per Equations 4.1 and 4.2). Note that the ALCR varies as a function of the TMP during operation, since a higher TMP results in a greater flow of water through the defect and thus a lower ALCR. In addition, these empirical relationships assume constant temperature, since changes in air or water temperature may change the functional relationships.

Although the CAM technique for determining the ALCR empirically is more labor-intensive than calculating the ALCR using a hydraulic model, the procedure does have several advantages. First, because the measurement is empirical, it is more accurate than calculations based on general hydraulic models. In addition, the CAM procedure does not rely on assumptions that are necessary to estimate the ALCR from hydraulic models, but instead facilitates direct determination of the ALCR based on measured air and water flows through a known defect. Another advantage of the CAM procedure is that it allows the ALCR to be easily recalculated for any operating TMP and direct integrity test pressure (assuming constant temperature).

D.2 Methodology

The following general procedure is provided as a guide for conducting the CAM technique for experimentally determining the ALCR. Note that although the CAM technique is described below for use with a single bench- or full-scale module for convenience in conducting the procedure, the resulting ALCR is scalable and independent of the size of the module or the integrity breach, as discussed in Appendix C (section C.1). Thus, the ALCR determined via this procedure would be applicable to an entire full-scale membrane unit.

1. Determine the baseline integrity test response for an integral bench- or full-scale membrane module. (See section 4.3.1.3 for a discussion of diffusive flow through the wetted pores of an integral membrane.)

2. For reference, measure the flow of water through the integral membrane module at various TMPs representative of the potential range of operating conditions.

3. Cut a known number of fibers (e.g., between 1 and 100). (Note that for many hollow-fiber membrane filtration systems, cutting a fiber at the point at which it enters the potting material represents the most conservative condition.)

4. Measure the water flow through the cut fiber(s) (i.e., Q_{water}) over the range of potential operating TMPs. One potential approach for determining Q_{water} is to compare the flow

through an integral membrane module (as determined in step 2, above) with that through the compromised module; the difference would represent the flow through the cut fiber(s) for each discrete TMP assessed. (Note that other approaches for determining Q_{water} may also be used.)

5. Develop an equation for a fitted curve that represents the water flow through the cut fiber(s) as a function of TMP.

6. Determine the minimum bubble point of the porous membrane material. (This information should generally be available from the manufacturer.)

7. Establish the direct integrity test pressure. As a general rule, the test pressure should be less than 80 percent of the bubble point pressure for the membrane and below the maximum TMP. However, the test pressure must be sufficient to meet the resolution requirement of the LT2ESWTR for the removal of *Cryptosporidium*, as expressed in Equations 4.1 and 4.2.

8. Measure the airflow from the cut fiber(s) at a variety of potential applied direct integrity test pressures. This may be particularly useful if the hydrostatic backpressure may vary between different direct integrity test applications.

9. If the diffusive flow (i.e., the baseline response, as determined in step 1) is significant (i.e., more than 5 percent of the total airflow) at the target test pressure to be utilized during normal operation (P_{test}), then a lower test pressure should be considered, if possible. (The test pressure must enable the direct integrity test to meet the resolution requirement.) Alternatively, the diffusive flow will have to be accounted for in determining the ALCR, as described in section 4.3.1.3.

10. Determine the ALCR using Equation D.1.

$$ALCR = \frac{Q_{air}}{Q_{water}}$$

Where:

ALCR	=	air-liquid conversion ratio (dimensionless)
Q_{air}	=	flow of air through the broken fiber(s) at the direct integrity test pressure (mL/min)
Q_{water}	=	flow of water through the broken fiber(s) at the reference TMP (mL/min)

Note that the reference TMP described in association with the variable Q_{water} above refers to the TMP that is used in the determination of the ALCR for the purpose of establishing the direct integrity test sensitivity for regulatory compliance, as described in section 4.3.1.2 and Appendix C. For example, the most conservative ALCR result would be generated from using the maximum anticipated TMP during normal operation.

11. Use Equation 4.7 to calculate the direct integrity test method sensitivity (i.e., LRV_{DIT}), incorporating the test result (Q_{air}) (either as directly measured with the diffusive airflow test or as converted from the pressure decay rate (ΔP_{test}) using Equation 4.8 if the pressure decay test is used) and the ALCR determined in the previous step, as described in Chapter 4. Note that the use of Equations 4.7 and 4.8 to calculate sensitivity require that the parameters Q_{air} and ΔP_{test}, respectively, represent the smallest integrity test response that can be reliably measured and associated with an integrity breach, as specified in section 4.3.1.2.

References

Adham, S., J.G. Jacangelo, and J.-M. Laine. 1995. Low-Pressure Membranes: Assessing Integrity. *Jour. AWWA*, 87(3):62.

Adham, S., Kuang-ping Chiu, Karl Gramith, and Joan Oppenheimer. 2005. *Development of a Microfiltration and Ultrafiltration Knowledge Base.* American Water Works Association Research Foundation (AwwaRF) Report 91059. Denver, Colo.: AwwaRF.

Alfrey, T. Jr. 1985. Structure-Property Relationships in Polymers. In *Applied Polymer Sciences*, ed. R.W. Tess and G.W. Poehlein, 2nd ed., pp. 241–252. American Chemical Society (ACS) Symposium Series 285. Washington DC. ACS

Allgeier, S.C. 2007. *Overview of Membrane Filtration Under the LT2ESWTR.* In *Proc. Membrane Technology Conference*, Tampa, Fla. Denver, Colo.: AWWA.

Alspach, B. 2007. *Advanced Concepts in Direct Integrity Testing Under LT2ESWTR.* In *Proc. Membrane Technology Conference*, Tampa, Fla. Denver, Colo.: AWWA.

Alspach, B., and Allgeier, S.C. 2003. Application of Ultrafiltration for Virus Removal Using the LT2ESWTR Regulatory Framework. In *Proc. Water Quality Technology Conference*, Philadelphia, Pa. Denver, Colo.: AWWA.

Alspach, B., and R.H. Sakaji. 2007. Comparing the Federal Regulatory Philosophies of Membrane Filtration Versus Conventional Media Filters. In *Proc. Membrane Technology Conference*, Tampa, Fla. Denver, Colo.: AWWA.

American Society for Testing and Materials (ASTM). 2003a. Practice A, Pressure Decay (PDT) and Vacuum Decay (VDT) Tests. ASTM D-6903. West Conshohocken, PA: ASTM International.

American Society for Testing and Materials (ASTM). 2003b. Standard Practice for Integrity Testing of Water Filtration Membrane Systems, ASTM D-6908-03. West Conshohocken, Pa.: ASTM International.

American Water Works Association Research Foundation (AwwaRF). 2000. Request for Proposal: Assessment and Development of Low-Pressure Membrane Integrity Monitoring Tools (RFP2681). Denver, Colo.: AwwaRF. Personal communication with then AwwaRF project manager, Dr. Jarka Popovicova

Anselme, C., and E.P. Jacobs. 1996. Ultrafiltration. In *Water Treatment Membrane Processes*, ed. J. Mallevialle, P.E. Odentaal, and M.R. Wiesner. New York: McGraw-Hill.

Askenaizer, D., R.H. Sakaji, and J.G. Jacangelo. 2005. Evaluation of California's Approach to Assignment of Pathogen Removal Credits and Membrane Acceptance as an Alternative Filtration Technology. In *Proc. Membrane Technology Conference*, Phoenix, Ariz. Denver, Colo.: AWWA.

Banerjee, A., K. Carlson, F. Marinelli, and K. DeNatale. 2000a. Monitoring Membrane Integrity with Laser Turbidity. In *Proc. AWWA Annual Conference*, Denver, Colo. Denver, Colo.: AWWA.

Banerjee, A., K. Carlson, and J. Lozier. 2000b. Monitoring Membrane Integrity Using Ultra High Sensitivity Laser Light. In *Proc. Water Quality Technology Conference,* Salt Lake City, Utah. Denver, Colo.: AWWA.

Banerjee, A., J. Lozier, and K. Carlson. 2001. An On-Line Multi-Sensor Membrane Filtration Permeate Water Quality Monitoring System. In *Proc. Membrane Technology Conference,* San Antonio, Texas. Denver, Colo.: AWWA.

Bersillon, Jean-Luc, and M.A. Thompson. 1996. Field Evaluation and Piloting. In *Water Treatment Membrane Processes,* ed. J. Mallevialle, P.E. Odentaal, and M.R. Wiesner. New York: McGraw-Hill.

Blevins, R.D. 1984. *Applied Fluid Dynamics Handbook,* New York. Van Nostrand Reinhold Company.

Booth, S., P. Carlson, and S. Kawamura. 2006. *Minimizing Operational Interruption During Filter Bed Surveillance,* American Water Works Association Research Foundation (AwwaRF) Report 91151, Denver, Colo.: AwwaRF.

Brown, L. 2004. *Continuous Imaging Fluid Particle Analysis—A Primer.* Fluid Imaging Technologies White Paper, http://www.fluidimaging.com/materials/literature.asp?category=1 (April 04 2008). Yarmouth, Me.: Fluid Imaging Technologies.

Brown, L. 2006. *Particle Imaging Understanding—A Primer.* Fluid Imaging Technologies, White Paper, http://www.fluidimaging.com/materials/literature.asp?category=1 (April 04 2008). Yarmouth, Me.: Fluid Imaging Technologies.

Carr, M., K.H. Carlson, and M. Sadar. 2003. Membrane Integrity Monitoring with Distributed Laser Turbidimeter. *Jour. AWWA,* 95(6):83–94.

Chellam, S., C.A. Serra, and M.R. Wiesner. 1998. Estimating the Cost of Integrated Membrane Systems. *Jour. AWWA,* 90(11):96–104.

Chemtrac Systems, Inc. 2003. Operation Manual for Particle Monitor PM2500XRD, Norcross, GA, Chemtrac systems, Inc.

Cheng J., E. l. Sheldon, L. Wu, A. Uribe, L. O. Gerrue, J. Carrino, M. Heller, and J. P. O'Connell 1998. Preparation and Hybridization Analysis of DNA/RNA from *E. coli* on Microfabricated Bioelectronic Chips. *Nature Biotechnology,* 16:541–546.

Cheryan, M. 1998. *Microfiltration and Ultrafiltration Handbook.* Lancaster, PA: Technomic Publishing Co.

Childress, A.E., and M. Elimelech. 1996. Effect of Solution Chemistry on the Surface Charge of Polymeric Reverse Osmosis and Nanofiltration Membranes. *Jour. Membrane Science,* 119(2):253–258.

Cleasby, J. L. and G. S. Logsdon 1999. Granular Bed and Precoat Filtration, in Water Quality and Treatment: A Handbook of Community Water Suppliers, fifth edit., edited by R. D. Letterman, McGraw-Hill, Inc., 1999, New York

Clements, J., G. Milton, J. Abrahamsson, and U. Eriksson. 2006. Ceramic Membranes and Coagulation for TOC Removal from Surface Water, in *Proc. of Water Quality Technology Conference*, Denver, CO., Denver, Colo.: AWWA.

Colvin, C., R. Brauer, N. DeNatale, and T. Scribner. 2001. Comparing Laser Turbidimetry with Conventional Methods for Monitoring MF and UF Membrane Integrity. In *Proc. Membrane Technology Conference*, San Antonio, Texas. Denver, Colo.: AWWA.

Corbitt, R. A. 1990. *Standard Handbook of Environmental Engineering*, edt. by R. A. Corbitt, p4.100, New York, McGraw-Hills, Inc.

Côté, P., I. Sutherland, N. Adams, and J. Cadera2003. Validation of Membrane Integrity Methods in a Challenge Test with *Bacillus subtilis*. In *Proc. Membrane Technology Conference*, Atlanta, Ga. Denver, Colo.: AWWA.

Crane Co. 1988. *Flow of Fluids Through Valves, Fittings, and Pipe*. Technical Paper 410. Stamford, Conn.: Crane Co.

DeVries, K. I. and R. J. Nuismer. 1985. Fracture Mechanics of Polymers, in *Applied Polymer Sciences*, 2nd Edt.,277–303. American Chemstry Society (ACS) Symposium Series, Vol. 285 Washington DC.: ACS

Dodge, B.T. 1944. *Chemical Engineering Thermodynamics*, pp. 349–351. New York: McGraw-Hill.

Douglas, I., D. Thomas, J. Guthmann, S. Russell, and S. Springthorpe 2004. Dynamic Particle Analysis: A New Technology for Optimizing Particle Removal in a Water Treatment Plant. In *Proc.,Water Quality Technology Conference,* San Antonio, Texas,. Denver, Colo.: AWWA.

Farahbakhsh, K., and D. Smith. 2003. Estimating Air Diffusion Contribution to Pressure Decay During Membrane Integrity Tests. In *Proc. Membrane Technology Conference*, Atlanta, Ga. Denver, Colo.: AWWA.

Franklin, B., R. Smith, and F. Knops. 2001. The Construction and Commissioning of a 24-mgd Ultrafiltration Plant with On-Line Integrity Testing to Monitor and Maintain the Barrier to *Cryptosporidium*-Sized Particles. In *Proc. Membrane Technology Conference,* San Antonio, Texas. Denver, Colo.: AWWA.

Garcia-Aleman, J., and J. Lozier. 2005. Managing Fiber Breakage: LRV Operations and Design Issues Under the LT2 Framework. In *Proc. Membrane Technology Conference*, Phoenix, Ariz. Denver, Colo.: AWWA.

Giles, R.V., J.B. Evett, and C. Liu. 1994. *Fluid Mechanics and Hydraulics,* 3rd ed. Schaum's Outline Series. New York: McGraw-Hill.

Glucina, K., J.M. Lainê, M. Chamant, and P. Simonie. 1999. Acoustic Sensor: A Novel Technique for Low-Pressure Membrane Integrity Testing. In *Proc. Membrane Technology Conference*, Long Beach, Calif. Denver, Colo.: AWWA.

Grant, D. 2004. Factors Affecting the Generation, Detection, and Removal of Microbubbles in Liquids. In *Proc. Semiconductor Pure Water and Chemicals Conference,* Santa Clara, Calif. Los Altos, Calif., SPWCC

Gregory, J. 1989. A Simple Particle Monitor for Low-Turbidity Waters. In *Proc. Membrane Technology Conference*, St. Louis, Mo. Denver, Colo.: AWWA.

Gregory, J. 2006. *Particles in Water: Properties and Processes.* London: IWA Publishing.

Hanfeld, J.L., and B.D. Dalke. 1989. General Purpose Polystyrene. In *Encyclopedia of Polymer Science and Engineering*, ed. J.I. Kroschwitz, vol. 16, pp. 62–71. New York: John Wiley and Sons.

Hansen, K. M., H. F. Ji, G. H. Wu, R. Datar, R. Cote, A. Majumdar, and T. Thundat. 2001. Cantilever-Based Optical Deflection Assay for Discrimination of DNA Single-Nucleotide Mismatches, *Anal. Chem.*, 73(7):1567–1571.

Hargesheimer E., C. Lewis, and N. McTigue. 2000. *Fundamentals of Drinking Water Particle Counting.* Denver, Colo.: AwwaRF.

Herschell, J. 2007. Survey of State/Territory Regulatory Approaches for Approval of Low-Pressure Membrane Systems. In *Proc. Membrane Technology Conference*, Tampa, Fla. Denver, Colo.: AWWA.

Hertzberg, R.W., and J.A. Manson. 1980. *Fatigue of Engineering Plastics.* New York: Academic Press.

Hong, S.K., J.S. Taylor, F. Miller, J. Rose, C. Gibbson, C. Owen, and W. Johnson. 1999. Removal of Microorganisms by MF Process: Correlation Between Membrane Integrity Testing Results and Microbial Removal Efficiency. In *Proc. Membrane Technology Conference*, Long Beach, Calif. Denver, Colo.: AWWA.

Jacobson, K.B. 2007. Biosensors and Other Medical and Environmental Probes. http://www.ornl.gov/info/ornlreview/rev29.3/text/biosens.htm (April 18 2008). Oak Ridge, Tenn. Oak Ridge National Laboratory.

Ji, H.F., R. Dabestani, G.M. Brown, and P.F. Britt. 2000. A Novel Self-Assembled Monolayer (SAM) Coated Microcantilever for Low Level Cesium Detection. *Chem. Commun.*, 2000(6):457–458.

Ji, H.F., T. Thundat, R. Dabestani, G. M. Brown, P. F. Britt, and P. V. Bonnesen. 2001. Ultrasensitive Detection of CrO42- Using a Microcantilever Sensor, *Anal. Chem.*, 73:1572–1576.

Johnson, W.T. 1997. Predicting Log Removal Performance of Membrane Systems Using In Situ Integrity Testing. In *Proc. AWWA Annual Conference*, Atlanta, Ga. Denver, Colo.: AWWA.

Jucker, C., and M. Clark. 1994. Adsorption of Aquatic Humic Substances on Hydrophobic Ultrafiltration Membranes. *Jour. Membrane Science*, 97:253–268.

Kothari, N., and E. St. Peter. 2000. Utility Perspective on Regulatory Approval for Microfiltration Treatment Facilities in Wisconsin. In *Proc. AWWA Annual Conference*, Denver, Colo. Denver, Colo.: AWWA.

Kruithof, J.C., P. Kamp, and H. Holmer. 2001. Membrane Integrity Monitoring at the UF/RO Heemskerk Plant In *Proc. Membrane Technology Conference*, San Antonio, Texas. Denver, Colo.: AWWA.

Kruithof, J.C., P.C. Kamp, H.C. Folmer, M.N. Nederlof, S.C. J.M. Van Hoof. 2000. Development of a Membrane Integrity Monitoring Strategy for the UF/RO Heemskerk Drinking Water Treatment Plant. In *Proc. Conference on Membranes in Drinking and Industrial Water Production*, Paris, France. London, UK. International Water Association.

Kumar, A., and R.K. Gupta. 1998. *Fundamentals of Polymers*. New York: McGraw-Hill.

Landsness, L.B. 2001. Accepting MF/UF Technology—Making the Final Cut. In *Proc. Membrane Technology Conference*, San Antonio, Texas. Denver, Colo.: AWWA.

Lerman, A. 1979. *Geochemical Process-Water and Sediments Environments*. New York: Wiley Interscience.

Liu, C. 1998. Developing Integrity Testing Procedures for Pall Microza™ MF and UF Modules. I. Diffusion Flow and Pressure Hold Methods. SLS Report 7426. Port Washington, N.Y.: Pall Corporation.

Liu, C. 2002. Assessment of Pressure-Hold Test of Hollow Fiber Membrane Systems in Drinking Water Filtration. SLS Report 8161CL. Port Washington, N.Y.: Pall Corporation.

Liu, C. 2003. Hydraulic Analysis of Flow Inside the Hollow Fibers. SLS Report 10696CL. Port Washington, N.Y.: Pall Corporation.

Liu, C. 2004a. Investigation of Integrity Failure Events at Highland Park Reservoir Microfiltration Plant, Pittsburgh, Pa. SLS Report 11845CL. Port Washington, NY: Pall Corporation.

Liu, C. 2004b. Sensitivity and Reliability of Membrane Integrity Testing. In *Proc. Membrane Technology Conference*, Quebec City, Quebec. Denver, Colo.: AWWA.

Liu, C. 2006. A Risk-Based Approach to Determine the Frequency of Integrity Testing for Drinking Water Membrane Systems. In *Proc. Membrane Technology Conference*, Denver, Colo. Denver, Colo.: AWWA.

Liu, C. 2007. Mechanical and Chemical Stabilities of Polymeric Membranes. In *Proc. Membrane Technology Conference*, November 2007, Charlotte, N.C. Denver, Colo.: AWWA.

Liu, C., and J.K. Schaefer. 2002. Defining and Quantifying Integrity: A Systematic Approach. In *Proc. Membrane Technology Conference*, Seattle, Wash. Denver, Colo.: AWWA.

Liu, C., and J.K. Schaefer. 2004. Regulatory Implications of the LT2ESWTR and Membrane Filtration Guidance Manual for Membranes in Drinking Water Treatment, *Proc. Texas Water 2004*, Arlington, TX, Austin, TX, Texas AWWA and WEA.

Liu, C., M. Fushijima, J. Hayes, and J. Moy. 1999. Finding a Needle in a Haystack: Membrane Integrity Testing in Drinking Water Applications. In *Proc. Membrane Technology Conference*, Tampa, Fla. Denver, Colo.: AWWA.

McCuin, R.M., and J.L. Clancy. 2005. Use of Microscopic Fluidic Imaging for Identification and Quantification of Organisms in Water. In *Proc. Membrane Technology Conference*, Quebec City, Quebec. Denver, Colo.: AWWA.

Moch, I. Jr., and D.J. Paulson. 2003. Establishing an Integrity Standard for Membrane Systems. In *Proc. International Desalination Association World Congress*, Paradise Island, Bahamas. Topsfield, MA. International Desalination Association

Thomas, D and P. Moor. 2004. Water Treatment Plant Particle Characterization Using Digital Image Technology, Proc. Canadian Water and Wastewater Association Annual Conference, Calgary, AB, Ottawa, ON. CWWA

Naismith, J. 2005. Five Years of Operational Experience at Plant C, a 7.8-mgd Microfiltration Water Treatment Plant. In *Proc. Membrane Technology Conference*, Phoenix, Ariz. Denver, Colo.: AWWA.

Panglisch, S., U. Deinert, W. Dautzenberg, O. Kiepke, and R. Gimbel 1998. Monitoring the Integrity of Capillary Membranes by Particle Counters. *Desalination*, 119:65–72.

Particle Measuring Systems (PMS). 1976 Background Noise in Optical Particle Counters: A Need to Establish Statistical Control. Technical Note 5185, http://www.pmeasuring .com/support/papers/particlemonitoring/liquid. (December 03 2007) Boulder, Colo.: PMS

Pashby, C., R.D. Barcus, and M.T. Sloan. 2000. Guidelines for Selecting an Optical Particle Counter. MetOne Application Note, http://www.iprocessmart.com/particlecounters/guidelines_for_selecting_an_opti.htm (December 07 2007). Grants Pass, Oregon, Pacific Scientific Instruments.

Petry, M., D. Thonny, J. P. Roux, P. Moulart, and V. Bonnelye. 2001. Lauss anne: Specific Design for a Challenge Project. In *Proc. Membrane Technology Conference*, San Antonio, Texas. Denver, Colo.: AWWA.

Phanttaranawik, J., A. G. Fane, and F. S. Wong, 2008. Novel Membrane-Based Sensor for On-line Membrane Integrity Monitoring, *J. Mem Science*, 323, 113 -124

Philippi, E.,G. W. Harrington, B. Lau, D. Thonas, and S. Russell. 2005. A Closer Look at Filter Effluent Particles Using Image-Based Particle Analysis. In *Proc. Membrane Technology Conference*, Denver, Colo. Denver, Colo.: AWWA.

Rabinski, G. and D. Thomas. 2004. Dynamic Digital Imagae Analysis: Emerging Technology for Particle Characterization, *Water Science and Technol.,* 50 (12): 19-26

Reich, L., and S.S. Stivala. 1971. *Elements of Polymer Degradation*. New York: McGraw-Hill.

Reilley-Matthews, B. 2007. Particle Imaging and Analysis Instrumentation Minimizes Taste and Odor Complaints. *Jour. AWWA*, 99(11):28–30.

Rogers, K., and M. Mascini. 2007. Biosensors for Analytical Monitoring, http://www.epa.gov/heasd/edrb/biochem/intro.htm. (April 03 2008) Washington DC, USEPA

Sadar, M. 2002a. Turbidity Instrumentation—An Overview of Today's Available Technology. *Turbidity and Other Sediment Surrogates Workshop,* Reno, Nev. Reston, Va.: USGS

Sadar, M. 2004. Introduction to Laser Nephelometry: An Alternative to Conventional Particulate Analysis Method. Hach Company L7044. http://www.hach.com/fmmimghach?/Colo.DE%3AL7044_02-057659%7C1. (December 16 2007) Loveland, Colo.: Hach Co.

Sadar, M., and L. Herrington. 2002. Evaluation of a New Continuous On-Line Integrity Testing Method for Membrane Filtration. In *Proc. Water Quality Technology Conference,* Seattle, Wash. Denver, Colo.: AWWA.

Sadar, M., L. Oxenford, M. Lichwardt, F. Watt, and J. Boswell. (2003a). Monitoring Membrane Integrity Across Four Microfiltration Pilot Plants Using Multiplexed Light-Scattering Technologies. In *Proc. Water Quality Technology Conference,* AWWA, November 2003, Philadelphia, Pa. Denver, Colo.: AWWA.

Sadar, M., V. Hart, D. Kolman, and L. Herrington. (2003b). Multiplexed Membrane Integrity Monitoring—From Pilot to Full-Scale Monitoring. In *Proc. Membrane Technology Conference,* AWWA, March 2003, Atlanta, Ga. Denver, Colo.: AWWA.

Sakaji, R.H., R. Haberman, and G. Yamamoto. 2001. The Multi-Barrier Concept: Is It Necessary? In *Proc. Membrane Technology Conference,* San Antonio, Texas. Denver, Colo.: AWWA.

Schideman, L., M. White, L. Landsness, M. Kosterman, and L. Rago. 2004. Challenges, Opportunities, and Solutions in Regulating the Design of Low-Pressure Membrane Water Treatment Plants. In *Proc. AWWA Annual Conference,* Orlando, Fla. Denver, Colo.: AWWA.

Schulze, J.C. 2001. The Texas Approach to Regulating MF/UF Drinking Water Applications. In *Proc. Membrane Technology Conference,* San Antonio, Texas. Denver, Colo.: AWWA.

Sepaniak, M., P. Datskos, N. Lavrik, and C. Tipple. 2002. Microcantilever Transducers: A New Approach in Sensor Technology. *Anal. Chem.,* 74:569A–575A.

Sethi S., G. Crozes, D. Hugaboom, B. Mi, J.M. Curt, and B. Mariñas. 2004. *Assessment and Development of Low-Pressure Membrane Integrity Monitoring Tools.* AwwaRF Report 91032. Denver, Colo.: AwwaRF.

Stumm, W., and J.J. Morgan (1996): *Aquatic Chemistry: Chemical Equilibria and Rates in Natural Waters,* 3rd Ed., New York. Wiley-Interscience Publication

Symons, G. 2006. Water Treatment Through the Ages. *Jour. AWWA,* 98(3)87.

Thomas, D. 2005. Application of Automatic Micro-Flow Imaging to Pathogen Detection and Enumeration. In *Proc. Joint Annual OWWA/OMWA Conference,* Ottawa, Ont. Ottawa, Ont. OWWA/OMWA

Thomas, D., and P. Moore. 2003. Particle Characterization Using Dynamic Digital Image Analysis. In *Proc. Water Quality Technology Conference,* Philadelphia, Pa. Denver, Colo.: AWWA.

Thominette, F., O. Farnault, E. Gaudichet-Maurin, C. Machinal, and J.-C. Shrotter. 2006. Aging of Polyethersulfone Ultrafiltration Membranes in Hypochlorite Treatment. *Desalination*, 200:7–8.

Thundat. T., G.Y. Chen, R.J. Warmack, D.P. Allison, and E.A. Wachter. 1995. Vapor Detection Using Resonating Microcantilevers, *Anal. Chem.,* 67:519–521.

Trimboli, P., J. Lozier, and W. Johnson. 2000. Demonstrating the Integrity of a Large-Scale Microfiltration Plant Using Bacillus Spore Challenge Test. In *Proc. Conference on Membranes in Drinking and Industrial Water Production*, Paris, France. London, UK. IWA

US Environmental Protection Agency (USEPA). 1989. "Surface Water Treatment Rule, Final Rule." *Federal Register*, 54: 27486 June 29 1989.

USEPA. 2000. "Stage 2 Microbial and Disinfection By-Products Federal Advisory Committee Agreement in Principle, Notice of Agreement in Principle." *Federal Register* 65:83015 (29 December 2000), p. 83015.

USEPA. 2001. *Low-Pressure Membrane Filtration for Pathogen Removal: Application, Implementation, and Regulatory Issues*. EPA 815-C-01-001. Washington, D.C.: USEPA.

USEPA. 2005. *Membrane Filtration Guidance Manual.* EPA 815-R-06-009. Washington, D.C.: USEPA.

USEPA. 2006a. "National Primary Drinking Water Regulations: Long-Term 2 Enhanced Surface Water Treatment Rule (LT2ESWTR), Final Rule." *Federal Register* 71:654 (January 5, 2006), p. 654–786.

USEPA. 2006b. "National Primary Drinking Water Regulations: Stage 2 Disinfectants and Disinfection Byproducts Rule: Final Rule". *Federal Register*, 71:388 (January 04 2006), p. 388–493

USEPA. 2006c. National Primary Drinking Water Regulations: Ground Water Rule: Final Rule". *Federal Register*, 71:65574 (November 08 2006), p. 65574–65660

Van Hoff, S.C.J.M., J.C. Kruithof, and P.C. Kamp. 2001. Development of a New On-Line Membrane Integrity Testing System. In *Proc. Membrane Technology Conference*, San Antonio, Texas. Denver, Colo.: AWWA.

Van Mises, R. 2004. *Mathematical Theory of Compressible Fluid Flow*. Mineola, N.Y.: Dover Publications.

Vashist, S.K. 2007. A Review of Microcantilever for Sensing Applications. http://www.azonano.com/details.asp?ArticleID=1927. (April 22, 2008)

Vickers, J. 1993. *Aspects of Integrity Testing and Module Construction of Microporous Membrane Filters.* Technical Paper. Rosedale, Md.: Memtec America Corporation.

Vieth, W.R. 1991. *Diffusion in and Through Polymers: Principles and Applications.*, New York. University Press.

Virginia Department of Health. 2006. Water—Design and Construction—Treatment Hollow Fiber, Positive Pressure Driven Microfiltration and Ultrafiltration Membrane Filtration Technology, Working Memo No. 880. Richmond, VA Virginia Department of Health

Webb, P.A. 2000. A Primer on Particle Sizing by Static Laser Light Scattering. Micromeritics Instrument Corp. Technical Workshop Series: Introduction to the Latest ANS/ISO Standard for Laser Particle Size Analysis. http://www.particletesting.com/docs/primer:particle_sizing_laser.pdf. (December 11, 2007) Norcross, Ga.: Micromeritics Instrument Corp.

Wu, G. H., R. H. Datar, K. M. Hansen, T. Thundat, R. J. Cote, and A. Majumdar. 2001. Bioassay of Prostate-Specific Antigen (PSA) Using Microcantilevers, *Nat. Biotechnol.* 19(9): 856—860.

Zondervan, E., B.H.L. Betlem, and B. Roffel. 2007. Accelerated Aging of PES UF Membranes. In *Proc. Membrane Technology Conference*, Tampa, Fla. Denver, Colo.: AWWA.

Index

Note: *f.* indicates figure; *t.* indicates table.

www.ingramcontent.com/pod-product-compliance
Lightning Source LLC
LaVergne TN
LVHW060314240926
848814LV00005B/31